Imed Mahfoudhi

Problèmes inverses de sources dans des équations de transport

Imed Mahfoudhi

Problèmes inverses de sources dans des équations de transport

Problèmes inverses de sources dans des équations de transport à coefficients variables

Presses Académiques Francophones

Impressum / Imprint

Bibliografische Information der Deutschen Nationalbibliothek: Die Deutsche Nationalbibliothek verzeichnet diese Publikation in der Deutschen Nationalbibliografie; detaillierte bibliografische Daten sind im Internet über http://dnb.d-nb.de abrufbar.

Alle in diesem Buch genannten Marken und Produktnamen unterliegen warenzeichen-, marken- oder patentrechtlichem Schutz bzw. sind Warenzeichen oder eingetragene Warenzeichen der jeweiligen Inhaber. Die Wiedergabe von Marken, Produktnamen, Gebrauchsnamen, Handelsnamen, Warenbezeichnungen u.s.w. in diesem Werk berechtigt auch ohne besondere Kennzeichnung nicht zu der Annahme, dass solche Namen im Sinne der Warenzeichen- und Markenschutzgesetzgebung als frei zu betrachten wären und daher von jedermann benutzt werden dürften.

Bibliographic information published by the Deutsche Nationalbibliothek: The Deutsche Nationalbibliothek lists this publication in the Deutsche Nationalbibliografie; detailed bibliographic data are available in the Internet at http://dnb.d-nb.de.

Any brand names and product names mentioned in this book are subject to trademark, brand or patent protection and are trademarks or registered trademarks of their respective holders. The use of brand names, product names, common names, trade names, product descriptions etc. even without a particular marking in this works is in no way to be construed to mean that such names may be regarded as unrestricted in respect of trademark and brand protection legislation and could thus be used by anyone.

Coverbild / Cover image: www.ingimage.com

Verlag / Publisher:
Presses Académiques Francophones
ist ein Imprint der / is a trademark of
OmniScriptum GmbH & Co. KG
Heinrich-Böcking-Str. 6-8, 66121 Saarbrücken, Deutschland / Germany
Email: info@presses-academiques.com

Herstellung: siehe letzte Seite /
Printed at: see last page
ISBN: 978-3-8381-4151-0

Copyright © 2014 OmniScriptum GmbH & Co. KG
Alle Rechte vorbehalten. / All rights reserved. Saarbrücken 2014

Résumé

Ce porte sur l'étude de quelques questions liées à l'identifiabilité et l'identification d'un problème inverse non-linéaire de source. Il s'agit de l'identification d'une source ponctuelle dépendante du temps constituant le second membre d'une équation de type advection-dispersion-réaction à coefficients variables. Dans le cas monodimensionnel, la souplesse du modèle stationnaire nous a permis de développer des réponses théoriques concernant le nombre des capteurs nécessaires et leurs emplacements permettant d'identifier la source recherchée d'une façon unique. Ces résultats nous ont beaucoup aidés à définir la ligne de conduite à suivre afin d'apporter des réponses similaires pour le modèle transitoire. Quant au modèle bidimensionnel transitoire, en utilisant quelques résultats de nulle contrôlabilité frontière et des mesures de l'état sur la frontière sortie et de son flux sur la frontière entrée du domaine étudié, nous avons établi un théorème d'identifiabilité et une méthode d'identification permettant de localiser les deux coordonnées de la position de la source recherchée comme étant l'unique solution d'un système non-linéaire de deux équations, et de transformer l'identification de sa fonction de débit en la résolution d'un problème de déconvolution. La dernière partie de ce livre discute la difficulté principale rencontrée dans ce genre de problèmes inverses à savoir la non identifiabilité d'une source dans sa forme abstraite, propose une alternative permettant de surmonter cette difficulté dans le cas particulier où le but est d'identifier le temps limite à partir duquel la source impliquée a cessé d'émettre, et donc ouvre la porte sur de nouveaux horizons.

Mots clés : Problèms Inverse de source ; Nulle contrôlabilité frontière ; Optimisation ; équation de diffusion-Advection-réaction ; Équation aux Dérivées Partielles ; Pollution des eaux de surfaces.

Table des matières

Table des matières

Introduction générale   9

I   Modélisation mathématique   13
    1   Introduction 13
         1.1   Principales sources de pollution des eaux de surface 13
         1.2   Conséquences de la pollution des eaux de surface 14
         1.3   Indices renseignant sur la présence d'une pollution 14
    2   Historique de la qualité des eaux de surface 16
    3   Modèles mathématiques 17
         3.1   Modèle simple $[DBO]$ 17
         3.2   Modèle couplé $[DBO] - [OD]$ 19
    4   But du travail : problème inverse de source 20

II   Identification of a point source in stationary $1D$ advection-dispersion-reaction equations with varying coefficients : detection of a pollution source   23
    1   Introduction 23
    2   Mathematical modelling and problem statement 24
    3   Identifiability 28
    4   Identification 30
    5   Numerical experiments 32
    6   Conclusion 36

Table des matières

III Inverse source problem in a one-dimensional evolution linear transport equation with spatially varying coefficients : application to surface water pollution 37
    1   Introduction 37
    2   Mathematical modelling and problem statement 38
    3   Identifiability 46
    4   Identification 48
        4.1   Step1 : Localization of the source position S 48
        4.2   Step2 : Recovery of the time-dependent intensity function λ 52
    5   Numerical experiments 55
        5.1   Undimensioned Problem 55
        5.2   Particular choice for the coefficients D, v and r 57
        5.3   Numerical tests and discussion 60
    6   Conclusion 64

IV Inverse source problem based on two dimensionless dispersion-current functions in 2D evolution transport equations with spatially varying coefficients 65
    1   Introduction 65
    2   Problem statement 67
    3   Identifiability 78
    4   Identification 80
        4.1   Localization of the sought source position 81
            4.1.1   Source localization procedure 83
            4.1.2   Computation of the HUM boundary control 84
        4.2   Identification of the time-dependent source intensity 88
    5   Numerical experiments 90

V Identification of the time active limit with lower and upper frame bounds of the total amount loaded by an unknown source in a 2D transport equation 99
    1   Introduction 99
    2   Mathematical modelling and prelimenary results 101
    3   Identifiability 107
    4   Identification 109
        4.1   Application to some frequently encountered pollution sources 113
    5   Numerical experiments 114
    6   Conclusion 122

VI Conclusion générale et perspectives 125

A Annexe 129
    1   Chapitre 3 129
    2   Chapitre 4 131

Bibliographie 133

Notations

Notation

Soit Ω un ouvert de $\mathbb{R}^n$ de frontière $\partial\Omega$ suffisamment régulière. Nous rappelons les définitions des ensembles suivants :

- $L^p(\Omega) := \left\{ u : \Omega \longrightarrow \mathbb{R} \text{ mesurable et } \int_\Omega |u(x)|^p \, dx < \infty \right\}.$

- $L^2(\Omega) := \left\{ u : \Omega \longrightarrow \mathbb{R} \text{ mesurable et } \int_\Omega |u(x)|^2 \, dx < \infty \right\}.$

- $L^2_\ell(0,l) := \left\{ u : (0,l) \longrightarrow \mathbb{R} \text{ mesurable et } \int_0^l u^2(x)\ell(x)dx < \infty \right\}.$

- $L^\infty(\Omega) := \{ u : \Omega \longrightarrow \mathbb{R} \text{ tel que } \exists\, c > 0 \;\; \|u(x)\| \leq c \;\; \text{p.p } x \in \Omega \}.$

- $\|u\|_{L^\infty(\Omega)} := sup\,ess\,\{\, |u(x)| \text{ p.p } x \in \Omega \,\} = \inf\{C \geq 0, \text{ tel que } |u(x)| \leq C \text{ p.p } x \in \Omega\}$

- $\|u\|_{L^2_\ell(0,l)} := \left(\int_0^l |u(x)|^2 \ell(x) \, dx \right)^{1/2}, \;\; \|u\|_{L^p(\Omega)} := \left(\int_\Omega |u|^p \, dx \right)^{1/p}.$

- $C^k(\Omega)$: L'ensemble des fonctions k fois dérivable dont la k-ième dérivée est continue.

- $C^k_0(\Omega) := \left\{ u \in C^k(\Omega) \text{ à support compact dans } \Omega \right\}.$

- $\mathcal{D}(\Omega) := $ L'ensemble des applications $C^\infty(\Omega)$ à support compact dans Ω.

- $H^1(\Omega) := \left\{ u \in L^2(\Omega) : \frac{\partial u}{\partial x_i} \in L^2(\Omega), \; \forall i = 1,...N \right\}$ L'espace de Hilbert d'ordre 1.

- $|\alpha| = \sum_{i=1}^N \alpha_i,$ Où $\alpha = (\alpha_1,...\alpha_N)$ multi-indice avec $\alpha_i \in \mathbb{N}$ pour tout $i = 1,...,N.$

- $D^\alpha u := \dfrac{\partial^{|\alpha|}}{\partial^{\alpha_1} \cdot\, ...\, \cdot \partial^{\alpha_N}} u$: α-ième dérivée partielle de u.

- $W^{k,p}(\Omega) := \left\{ u \in L^p(\Omega) \text{ tel que } \forall \alpha \in \mathbb{N}^N, \; |\alpha| \leq k, \;\; D^\alpha u \in L^p(\Omega) \right\}.$

Notations

- $W^{k,\infty}(\Omega) := \left\{ u \in L^\infty(\Omega) \text{ tel que } \forall \alpha \in \mathbb{N}^N, \ |\alpha| \leq k, \ D^\alpha u \in L^\infty(\Omega) \right\}$.
- $H^k(\Omega) := W^{k,2}(\Omega) := \left\{ u \in L^2(\Omega) \text{ tel que } \forall \alpha \in \mathbb{N}^N, \ |\alpha| \leq k, \ D^\alpha u \in L^2(\Omega) \right\}$.
- $H_0^k(\Omega) := \left\{ u \in H^k(\Omega) \text{ tel que } u = 0 \text{ sur } \partial\Omega \right\}$: l'adhérence de $\mathcal{D}(\Omega)$ dans $H^k(\Omega)$.
- $H^{-k}(\Omega)$: L'espace des formes linéaires continues sur $H_0^k(\Omega)$, telles qu'il existe $C > 0$ pour laquelle $\forall \phi \in \mathcal{D}(\Omega), |<u, \phi>| \leq C\|\phi\|_{H^k(\Omega)}, \ <u,v> := \sum_{|\alpha| \leq k} \int_\Omega \partial^\alpha u . \partial^\alpha v dx$.
- $\|u\|_{H^k(\Omega)} := \sqrt{<u,u>}$ et $\|u\|_{H^{-k}(\Omega)} := \min\left\{ \frac{|<u,\phi>|}{\|\phi\|_{H^k(\Omega)}}, \ \phi \in \mathcal{D}(\Omega) \right\}$.
- $L^2((0,T); \mathcal{H}) := \left\{ u : [0,T] \longrightarrow \mathcal{H} : t \longrightarrow u(t) \text{ mesurable tel que } \int_0^T \|u(t)\|_{\mathcal{H}}^2 dt \ < +\infty \right\}$.
- $C^0((0,T); \mathcal{H}) := \{ u : [0,T] \longrightarrow \mathcal{H} : t \longrightarrow \|u(t)\|_{\mathcal{H}} \text{ continue}\}$.
- $\|V\|_2 := \sqrt{v_1^2 + \cdots + v_N^2}$ et $\|V\|_\infty = max\{|v_i|, \ i = 1, ..., N \ \}$.
- $\nabla u(x) = (\frac{\partial u}{\partial x_1}(x), \cdots, \frac{\partial u}{\partial x_N}(x))$: gradient de u.
- $\Delta u(x) = \sum_{i=1}^N \frac{\partial^2 u}{\partial x_i^2}$: laplacien de u.
- $\text{div}(V(x)) = (\frac{\partial v_1}{\partial x_1}(x) + \cdots + \frac{\partial v_N}{\partial x_N}(x))$ où $V(x) = (v_1(x), \cdots, v_N(x))$ l'opérateur de divergence.
- $\text{rot}(V(x)) = \begin{bmatrix} \frac{\partial v_3}{\partial x_2}(x) - \frac{\partial v_2}{\partial x_3}(x) \\ \frac{\partial v_1}{\partial x_3}(x) - \frac{\partial v_3}{\partial x_1}(x) \\ \frac{\partial v_2}{\partial x_1}(x) - \frac{\partial v_1}{\partial x_2}(x) \end{bmatrix}$ où $V(x) = (v_1(x), v_2(x), v_3(x))$ l'opérateur rotationnel
- $\det(A)$: Le déterminant de la matrice A.
- $\liminf_{n \to \infty} u_n = sup_n\{ \ inf_k\{ \ |u_k|, \ k \geq n \ \}, \ n \in \mathbb{N} \ \}$.
- ν : La normale extérieure au bord $\partial\Omega$ de Ω.
- χ_D : Fonction indicatrice de D tel que $\chi_D(x) = 1$ si $x \in D$ et 0 sinon.
- δ_a : La masse de Dirac au point a.
- $\mathcal{H}_a$: La fonction de Heaviside $\mathcal{H}_a(x-a) = 1 \ si \ x \geq a$ et 0 sinon.

Introduction générale

La modélisation mathématique est un outil principal pour la résolution des problèmes physiques. En effet, dans plusieurs domaines de la vie courante tels que la biologie, la géophysique, la chimie, etc ... l'évolution spatiale et/ou temporelle des phénomènes physiques peut être modélisée par des équations aux dérivées partielles/ordinaires. Étant donnés tous les paramètres intervenants dans le modèle mathématique utilisé, la résolution numérique de ces équations permet de mieux comprendre les phénomènes en jeu et même de prévoir un comportement anormal afin d'éviter des conséquences indésirables.

Par ailleurs, l'évolution scientifique notamment chimique et biologique a contribué à l'émergence d'un nouveau type de problèmes. Il s'agit des problèmes où quelques paramètres du modèle mathématique utilisé sont inconnus et en compensation de cette information manquante, des données sur les conséquences engendrées par l'action des ces inconnus sont disponibles. Ce nouveau genre de problèmes, appelé communément *Problème inverse* consiste en l'identification des paramètres inconnus à partir des données disponibles. De point de vue mathématique, ce type des problèmes soulève trois questions majeures : est-ce que les données disponibles permettent d'identifier les paramètres inconnus de façon unique ? comment peut-on remonter à partir de ces données aux inconnus ? est-ce que la présence de petites perturbations sur les données pourrait donner des paramètres tout a fait différents ? ensuite, la théorie des problèmes inverses a été principalement fondée autour de ces trois questions. Hadamard en 1923 dans [27] postule qu'un problème inverse est bien posé s'il satisfait les trois conditions suivantes :

1. la solution existe (Identification)
2. L'unicité de la solution (Identifiabilité)
3. la dépendance est continue par rapport aux données (Stabilité)

Table des matières

En effet, il est facile d'imaginer que la même conséquence puisse provenir de deux causes différentes ce qui souligne la première question que nous venons d'évoquer. Donc, mathématiquement pour les mêmes données il pourrait correspondre plusieurs paramètres alors que généralement un seul parmi eux est associé au problème physique étudié. Par exemple, "Peut-on entendre la forme d'un tambour ?" c'est la question que le mathématicien américain d'origine polonaise Mark Kac s'est posé. Cette question pourrait s'écrire mathématiquement de la manière suivante : connaissant la liste des fréquences propres, peut-on déduire la forme du cadre d'un tambour ? il s'agit d'un problème inverse qui consiste à identifier le bord du tambour comme une courbe simple fermée $\mathcal{C}$ du plan. Après 36 ans de recherche, la réponse à cette question a été énoncée en 1992 par les mathématiciens américains *C. Gordon, D. Webb* et *S. Wolpert*. : non, on ne peut pas identifier la forme d'un tambour à partir du son émis. La raison est que l'analyse des vibrations et l'acoustique montre qu'on peut toujours trouver deux formes différentes émettant des vibrations similaires.

Néanmoins, fort de ce caractère pratique permettant de remonter à partir des conséquences aux effets inconnus, nous avons vu durant les deux dernières décennies les problèmes inverses impliqués dans plusieurs domaines de la science : en médecine, par exemple, l'étude de l'électro-encéphalographie (EEG) [18, 22] ou aussi la restauration des activités du coeur humain à partir des mesures de potentiel sur le corps [64]. En géophysique, par exemple, l'identification de l'hypocentre (foyer) d'un séisme à partir des ondes ressenties ainsi que le suivi de la dynamique des séismes [40, 64]. En environnement, par exemple, l'identification des sources de pollution dans l'atmosphère, dans les eaux de surface [35, 36]. Nous citons aussi juste à titre indicatif quelques autres domaines qui ont tiré profit de l'application des problèmes inverses :

Chimie : identification des constantes de réactions chimiques.
Traitement d'image : restrauration des images floues.
Biomédical : localisation de sources épileptiques en EEG. Identification des gènes
défectueux. identification des cellules cancéreuses.
Énergie : identification des poches souterraines de pétrole.
Biologie : identification et suivi des sources d'épidémie dans des troupeaux.

Ce livre porte sur l'étude d'un problème inverse non-linéaire de source qui consiste en l'identification des sources dans des équations de type advection-dispersion-réaction appelées aussi équations de transport. Dans la pratique, ce type d'équations peut couvrir un large spectre d'applications allant de la biologie à l'étude de l'environnement en passant par la chimie, la médecine, etc ... D'ailleurs, une motivation pour ce livre est une application environnementale qui consiste en l'identification des sources de pollution dans des eaux de surface. Ceci nous amène à préciser l'organisation de ce manuscrit. En effet, ce livre comporte trois grandes parties dont le contenu est décrit comme suit :

Première Partie :

La première partie de ce livre consiste en un chapitre destiné à introduire une motivation de ce travail et à déboucher sur le modèle mathématique à considérer dans la suite. En effet, lors de ce chapitre nous partons du problème physique à étudier pour en déduire les différents modèles mathématiques associés. Nous discutons les avantages et les inconvénients de chaque modèle et finissons ce chapitre par préciser le modèle retenu pour la suite de notre étude.

Deuxième Partie :

Dans la deuxième partie, nous commençons l'étude du problème inverse de source dans le cas d'un modèle mathématique monodimensionnel. Nous entamons cette partie par un chapitre consacré à l'étude du régime stationnaire. En effet, ce régime nous offre beaucoup de souplesse pour développer des réponses théoriques aux deux questions principales évoquées au début de cette introduction : quelles sont les données qui permettent d'identifier la source recherchée de façon unique ? étant données ces mesures, comment peut-on déterminer la source inconnue ?

Comme nous allons le découvrir dans cette partie, les résultats obtenus dans le premier chapitre concernant le modèle stationnaire vont nous donner beaucoup de visibilité sur ce que pourrait se passer dans le cas d'un régime transitoire et donc, vont nous aider à définir la ligne de conduite à suivre afin d'établir l'identifiabilité et développer une méthode d'identification. Cela constitue le sujet du deuxième chapitre de cette partie.

Troisième Partie :

Quant à la dernière partie de ce livre, elle est composée de deux chapitres portant sur l'étude du problème inverse de source dans le cas d'un modèle mathématique bidimensionnel. Bien que le premier chapitre représente plutôt une continuité des deux chapitres de la deuxième partie, le second chapitre de cette dernière partie soulève une nouvelle question importante concernant l'identifiabilité d'une source dans sa forme abstraite et constitue donc une ouverture sur d'autres horizons.

Nous finissons le manuscrit par une conclusion générale permettant de résumer tous les résultats obtenus dans le cadre de ce livre, de rappeler les questions qui restent ouvertes et de donner quelques perspectives.

Chapitre 1

Chapitre I

Modélisation mathématique

1 Introduction

Ce livre porte sur l'étude d'un problème inverse non-linéaire de source. Une motivation pour ce travail est un problème environnemental qui consiste en l'identification des sources de pollution dans des eaux de surface. En fait, l'introduction de la matière organique dans l'eau d'une rivière, par exemple, incite les micro-organismes vivants dans cette rivière à particiiper au processus d'auto-épuration naturelle c.-à-d. à la dégradation de la matière organique introduite. Ce processus, généralement connu sous le nom d'oxydation, nécessite une grande consomation d'oxygène dissous dans l'eau. Par conséquent, plusieurs espèces vivantes dans la rivière peuvent mourir asphyxiées par manque d'oxygène et c'est d'ailleurs dans ce sens que la pollution constitue une grande menace pour la faune et la flore. En outre, les méfaits d'un tel phénomène peuvent aussi affecter toutes les activités de la vie humaine autour de la rivière : odeurs désagréables, pollution pour les stations de pompage d'eau potable, pollution pour les activités d'agriculture, etc ....

L'identification des sources de pollution dans une rivière permettrait non seulement d'améliorer la qualité de la vie humaine organisée autour de la rivière mais également de préserver la diversité du milieu aquatique par le fait d'apporter des réponses appropriées à une pollution identifiée.

1.1 Principales sources de pollution des eaux de surface

Les sources de pollution des eaux de surface sont classées essentiellement en trois catégories différentes :

Pollution domestique : ce type de pollution est en étroite relation avec les besoins quotidiens des ménages, et résulte de la collecte des eaux usées domestiques et de l'assainissement

Chapitre I. Modélisation mathématique

urbain. Il s'agit généralement d'une pollution composée de matières organiques et de produits chimiques (des détergents, des pesticides et rejets des hôpitaux).

Pollution agricole : les rejets provenant de l'agriculture constituent essentiellement une pollution bactériologique engendrée par la dégradation biologique de certaines végétations ainsi que les fertilisants naturels et/ou industriels (les engrais chimiques de nitrates et de phosphates). En cas d'excès d'emploi, ces produits peuvent altérer la qualité des eaux d'irrigation qui viennent par ruissellement et/ou infiltrations contaminer les cours d'eau et les eaux souterraines.

Pollution industrielle : cette catégorie de pollution est autant diversifiée que l'industrie même. La pollution industrielle est issue de l'ensemble des rejets industriels dans la nature sous forme d'aérosols (fumées ou poussières), dans les eaux usées et/ou déchets naturels industriels. Les dégradations biologiques de certains déchets et le dépôt des aérosols entraînent la pollution chimique et/ou organique des eaux de surface et souterraines.

1.2 Conséquences de la pollution des eaux de surface

L'apparition d'une pollution dans un milieu aquatique le déséquilibre et pourrait modifier sérieusement la nature de sa faune et sa flore. En effet, plusieurs espèces vivants peuvent en mourir asphyxié par manque d'oxygène ce qui réduirait la capacité du milieu aquatique à assurer le processus d'auto-épuration naturelle. En outre, la présence de la pollution dans les eaux d'une rivière, par exemple, affecte aussi les activités humaine organisées autour de cette rivière : odeurs désagréables, arrêt des stations de pompage d'eaux potable, arrêt de certaines activités d'agriculture, etc ....

1.3 Indices renseignant sur la présence d'une pollution

L'oxygène est un élément indispensable pour tout organisme vivant et tout milieu présentant un manque d'oxygène pourrait rapidement se transformer en un milieu sans vie. En partant de ce principe, la quantité disponible d'oxygène dissous dans l'eau constitue un indice majeur sur la qualité du milieu aquatique étudié. Dans la littérature, nous trouvons essentiellement trois indices pouvant renseigner sur la présence d'une pollution dans un milieu aquatique. Ces indices sont les suivants :

Demande Biologique en Oxygène [DBO] : représente la quantité d'oxygène consommée par les micro-organismes afin d'assurer la dégradation de la matière organique introduite dans l'eau. Le temps de digestion retenu est de 5 jours d'où l'appellation DBO_5. La mesure

I.1 Introduction

devra tenir compte de la température d'échantillon à 20C. Cette mesure s'exprime en milligramme d'oxygène par litre d'effluent et se calcule par la différence entre la mesure d'oxygène dans l'effluent à l'instant 0 et celle après 5 jours.

Oxygène Dissous [OD] : c'est la quantité d'oxygène présente dans l'eau qui est nécessaire à la vie aquatique et à l'oxydation des matières organiques. Ces dernières sont essentielles à la vie aquatique en tant que nourriture et sont mesurées en nombre de milligrammes d'oxygène (O_2) par litre d'eau (mg/L). L'eau est dite polluée lorsque la concentration en oxygène dissous est faible. On peut aussi mesurer le déficit en oxygène comme la différence entre la concentration de saturation et la concentration actuelle d'oxygène dissous dans l'eau.

Demande Chimique en Oxygène [DCO] : c'est un paramètre important pour la qualité d'eau. Elle représente la quantité d'oxygène nécessaire pour oxyder la matière organique contenue dans l'échantillon. Cette valeur est obtenue en faisant réagir des échantillons d'eau avec un oxydant puissant (le bichromate de potassium) et s'exprime en milligramme d'oxygène par litre d'eau.

Par qualité des eaux de surface, nous entendons les propriétés physiques, chimiques et biologiques. Elles peuvent être évaluées en mesurant la quantité de matière organique en suspension, de produits azotés et phosphorés contenus dans les eaux [9, 25]

Matières organiques : [25, 28] c'est la matière carbonée produite en général par des êtres vivants, animaux et micro-organismes. Il s'agit d'un ensemble de substances organiques dont la dégradation implique une consommation de l'oxygène dissous dans l'eau, ce qui a des conséquences directes sur la vie aquatique, dans ce sens constitue une pollution. Ces substances sont contenues dans les rejets, des eaux usées (zone urbaine) et des industries...

Matières azotées [9, 25, 28] : les différentes formes de l'azote mesurées dans l'eau de surface (rivières, fleuves, lacs, etc.), sont essentiellement de l'ammonium NH_4, des nitrites NO_2 et des nitrates NO_3. La présence de l'azote ammoniacal dans les eaux de surface peut être due à la décomposition des matières organiques azotées contenues dans les végétaux des algues, ou des rejets d'origine urbains ou industriels. Certaines industries (la production de glace, industrie du textile, etc.) contribuent à l'augmentation des concentrations en azotes d'ammoniaque.

Les nitrites constituent une étape importante dans la métabolisation des composés azotés. Ils s'insèrent dans le cycle de l'azote entre l'ammoniaque et les nitrates. Leur présence est donc due soit à l'oxydation bactérienne de l'ammoniaque, soit à la réduction des nitrates.

Les nitrates constituent le stade final de l'oxydation de l'azote suivant les réactions suivantes :

Chapitre I. Modélisation mathématique

$$NH_4 + \tfrac{3}{2}O_2 \longrightarrow NO_2 + H_2O + 2H$$
$$NO_2 + \tfrac{1}{2}O_2 \longrightarrow NO_3$$

Par ailleurs, l'activité humaine constitue une source incontestable de nitrates. Les apports proviennent essentiellement du lessivage des engrais et de l'azote reminéralisé sur les zones de culture, ainsi que des eaux usées et parfois des industries.

Matières phosphorées[28] : les eaux usées contiennent du phosphore sous forme organique (rejet humain) ou sous forme de phosphate (lessives). Les phosphates comme les nitrates contribuent au phénomène d'eutrophisation des cours d'eau et des lacs. Cela se traduit par l'enrichissement de l'eau par des matières fertilisantes, en particulier par des composés d'azote et de phosphore, qui à température élevée, accélèrent la croissance des algues végétales, ce qui entraîne une dé-oxygénation des eaux.

2 Historique de la qualité des eaux de surface

Au début des années 1920, la commission du fleuve Ohio aux États-Unis a commencé une étude intensive sur les sources de pollution et leurs impacts sur l'approvisionnement en eau domestique. De cette étude a émergé le premier modèle mathématique décrivant un environnement aquatique. C'est le modèle de Streeter-Phleps (1925) [9, 26, 54]. Ce modèle est fondé sur l'idée que la variation du taux de l'oxygène dissous dans une rivière pourrait être représenté par deux processus ; le premier étant un déficit dû à l'oxydation biochimique de la matière organique ; le second, un gain en oxygène provenant de l'air atmosphérique.

Dans le début des années 1960, plusieurs critiques sur le modèle de Streeter-Phleps ont été formulées. Ces dernières portent surtout sur le fait que ce modèle ne prend pas en compte tous les processus physiques se déroulant dans une rivière. En particulier, il ne tient compte ni de la contribution des algues sur le taux de déficit en oxygène (les algues consomment de l'oxygène dissous par respiration et elle produisent de l'oxygène par photosynthèse), ni de l'effet des facteurs climatiques sur les coefficients hydrauliques, ... Diverses modifications y ont été apportées [26] afin de tenir compte au mieux de la réalité. Elles ont été apportées essentiellement par Thomas (1948), O'Connor (1962), Camp (1963), Dobbins (1964). Au début des années 1970, commencent à apparaître les premières versions automatisées d'un modèle mathématique sur les traitement des eaux de surface. En effet, Board (1970) [26, 39, 54] a développé le modèle

$QUALI$ qui permet de simuler les variations de la $[DBO]$ et de la $[DO]$ (concentration de déficit en oxygène dissous) en fonction de plusieurs facteurs tels que la température de l'eau, le changement de flux,... . Ce modèle a été étendu, dans un premier temps, par Orlob (1982) [26, 54] afin de pouvoir tenir compte des processus qui peuvent se produire en présence d'azote.

Ensuite, Brown et Barwell (1987)[9, 26, 54] y ont intégré des équations permettant de modéliser la contribution des algues et de tenir compte des cycles de phosphore qui se produisent. Ce dernier modèle a été nommé $QUAL2$.

3 Modèles mathématiques

3.1 Modèle simple $[DBO]$

En se référant à l'historique établi dans la section précédente sur l'étude de la qualité des eaux de surface, il vient que l'évolution de la concentration de DBO, notée ici u, dans une partie de rivière modélisée par un ouvert Ω de $\mathbb{R}^n$ et surveillée pendant un temps T est régie dans $\Omega \times (0,T)$ par une équation de type advection-dispersion-réaction [1, 9, 25, 26, 54] :

$$\partial_t u - \operatorname{div}\left(D\nabla u - Vu\right) + R(u) = F \tag{I.1}$$

Où V, D, R et F désignent, respectivement, le champ de vitesse d'écoulement, le tenseur de dispersion, le terme de réaction et la source de pollution. En outre, comme le fluide étudié est l'eau d'une rivière, le modèle mathématique introduit en (I.1) peut être simplifié davantage. Pour cela, nous rappelons la définition suivante :

Définition : un fluide est dit incompressible lorsque son volume demeure quasiment constant sous l'action d'une pression externe.

En fait, la compressibilité d'un fluide mesure la variation du volume par rapport au volume total de ce fluide lorsqu'il est soumis à une pression extérieure. Par exemple [63], si on bouche l'orifice de sortie d'une pompe à vélo et que l'on pousse sur la pompe, on voit que l'on peut comprimer l'air contenu à l'intérieur. En revanche, si on refait la même expérience avec de l'eau à l'intérieur, on ne pourrait quasiment plus déplacer la pompe. C'est parce que la compressibilité de l'eau et d'ailleurs de tous les liquides est très faible. Pour cela et afin de simplifier les équations de la mécanique des fluides, on considère souvent que les liquides sont incompressibles. Mathématiquement, cela signifie que la masse volumique ϱ d'un tel fluide est supposée constante

Chapitre I. Modélisation mathématique

et donc l'équation de conservation des masses prend alors la forme suivante :

$$\partial_t \varrho + \text{div}(\varrho V) = 0 \quad \Leftrightarrow \quad \text{div}(V) = 0 \qquad (I.2)$$

Par conséquent, en se référant à (I.1) et (I.2) il vient que l'évolution de la [DBO] est régie par le modèle mathématique suivant :

$$\partial_t u - \text{div}(D\nabla u) + V\nabla u + R(u) = F \quad \text{dans } \Omega \times (0, T) \qquad (I.3)$$

u : La concentration de la [DBO] (mg/L)
D : Tenseur de dispersion (m^2/s)
V : Champ de vitesse d'écoulement (m/s)
R : Terme de réaction (s^{-1})
F : Source de pollution $(mgL^{-1}s^{-1})$

En ce qui concerne les coefficients D, V et R impliqués dans le modèle mathématique obtenu en (I.3), Dobbins en 1964 [15] a considéré le terme de réaction linéaire suivant :

$$R([DBO]) = (K_d + K_s)[DBO]$$

Où K_d et K_s sont, respectivement, les taux de dé-oxygénation et de sédimentation. Quant au tenseur de dispersion D, Bear et al. reportaient dans [5] que la dispersion hydrodynamique se produit généralement comme la conséquence des deux phénomènes suivants : la diffusion moléculaire résultant du mouvement aléatoire des molécules et la dispersion mécanique causée par des vitesses non-uniformes d'écoulement. La somme de ces deux phénomènes définie ce qu'on appelle tenseur de dispersion hydrodynamique. Donc, on a

$$D = D_M \mathbf{I} + \begin{bmatrix} D_1 & D_0 \\ D_0 & D_2 \end{bmatrix} \qquad (I.4)$$

Où $D_M > 0$ est un nombre positif représentant la diffusion moléculaire, $\mathbf{I}$ est la 2×2 matrice d'identité et les coefficients $D_{i=0,1,2}$ sont tels que [48, 49] :

I.3 Modèles mathématiques

$$D_1 = \frac{D_L V_1^2 + D_T V_2^2}{\|V\|_2^2}, \qquad D_0 = \frac{V_1 V_2 (D_L - D_T)}{\|V\|_2^2} \qquad \text{and} \qquad D_2 = \frac{D_L V_2^2 + D_T V_1^2}{\|V\|_2^2} \qquad (I.5)$$

Ici, $\|V\|_2 = \sqrt{V_1^2 + V_2^2}$ et D_T, D_L désignent les coefficients de dispersion transversale et longitudinale tels que $0 \leq D_T < D_L$. En outre, à partir de (I.4)-(I.5) nous réécrivons le tenseur de dispersion D comme suit :

$$D = \left(D_M + D_T\right)\mathbf{I} + \frac{D_L - D_T}{\|V\|_2^2} V V^\top \implies (D_M + D_T)\|X\|_2^2 \leq DX.X \leq (D_M + D_L)\|X\|_2^2 \ (I.6)$$

Pour tout $X \in \mathbb{R}^2$. L'implication dans (I.6) montre que la matrice D est uniformément elliptique et bornée dans Ω.

3.2 Modèle couplé $[DBO] - [OD]$

Dobbins dans [15] a introduit également un deuxième modèle mathématique sur la modélisation des indices renseignant sur la présence d'une pollution. Ce modèle, appelé modèle couplé, établi un lien entre la concentration de la DBO et celle du déficit en oxygène dissous dans l'eau, noté ici v, de la manière suivante :

$$\begin{aligned} \partial_t u - \mathrm{div}(D\nabla u) + V\nabla u + R(u) &= F \\ \partial_t v - \mathrm{div}(D\nabla v) + V\nabla v + R(v) &= Ru \end{aligned} \qquad (I.7)$$

Le terme de réaction de $[DBO]$ est identique à celui du premier modèle mais celui du déficit en oxygène dissous est donné par :

$$R([O_2]) = K_a([O_2]_s - [O_2]) - (K_d + K_s)[DBO]$$

Où k_a est le taux de ré-aération. Ce modèle est jugé plus réaliste sur le plan physico-chimique car il permet de décrire l'équilibre de ré-oxygénation et par ailleurs, plus praticable vis à vis des méthodes inverses car il explicite un paramètre plus mesurable en l'occurrence la concentration en oxygène dissous.

Un troisième modèle, plus complet que les deux précédents, a été développé par l'Agence

Chapitre I. Modélisation mathématique

de l'eau Seine Normandie, le Syndicat des Eaux de l'Ile de France et la compagnie générale des Eaux. En plus de la DBO et de l'oxygène dissout, ce modèle tient compte aussi de l'ammonium $NH4$ et des nitrates $NO3$. Le terme de réaction de $[DBO]$ est encore inchangé, alors que ceux de l'oxygène dissous, de l'ammonium et des nitrates sont données comme suit :

$$R([O_2]) = K_a([O_2]_s - [O_2]) - (K_d + K_s)[DBO] - 3.56 K_1[NH_4]$$
$$R([NO_3]) = -3.44 K_1[NH_4] \quad \text{et} \quad R([NH_4]) = K_2[NH_4]$$

D'autres modèles plus complexes que les précédents, considérant le couplage de plusieurs équations, ont été développés par l'US Environnant Protection Agency (1973) dans le programme $QUALII$ [9].

4 But du travail : problème inverse de source

Une motivation pour ce livre est l'identification des sources de pollution dans des eaux de surface. En se référant à la modélisation mathématique des indices renseignant sur la présence d'une source de pollution établie dans la section précédente, ce problème inverse de source peut s'écrire mathématiquement de la façon suivante : identifier la source de pollution F en utilisant l'un des deux modèles ci-dessous

– **Modèle simple** $[DBO]$: avec des données disponibles sur u.
– **Modèle couplé** $[DBO] - [OD]$: avec des données disponibles sur v.

Bien que le modèle simple présente l'avantage principal consistant en l'identification de la source F à partir des données sur les conséquences directes engendrées par la présence de cette source, il possède l'inconvénient de ne pas pouvoir fournir les résultats d'identification immédiatement vu que les données sur la $[DBO]$ nécessitent 5 jours de traitement au laboratoire afin d'être disponibles. En revanche, bien que le modèle couplé constitue un remède à l'inconvénient du modèle simple vu que les données sur la $[OD]$ sont généralement disponibles tout de suite, il pourrait être mathématiquement moins pratique et instable en raison de son caractère indirect c.-à-d. identifier F à partir des données sur les conséquences des conséquences.

Dans ce livre, nous nous concentrons sur l'étude du problème inverse de source en n'utilisant que le modèle simple et des données sur la $[DBO]$. Par ailleurs, il est bien connu dans la littérature que nous ne pouvons pas identifier de façon unique à partir des données frontières une source dans sa forme abstraite. Pour illustrer cette remarque, nous présentons l'exemple

I.4 But du travail : problème inverse de source

standard [16] suivant : soient $\Omega \subset \mathbb{R}^n$ avec $n = 2, 3$, $f \in \mathcal{D}(\Omega)$ et $g = -\Delta f$. Alors, la variable $u(x,t) = tf(x)$ résout le problème suivant :

$$\begin{cases} \partial_t u - \Delta u = f(x) + tg(x) & \text{dans} \quad \Omega \times (0,T) \\ u(x,0) = 0 & \text{dans} \quad \Omega \\ u(x,t) = \frac{\partial u}{\partial n}(x,t) = 0 & \text{sur} \quad \partial\Omega \times (0,T) \end{cases}$$

Ceci montre que toutes sources F_1 et F_2 qui sont telles que $F_1 - F_2 = f + tg$ produisent les mêmes données sur le bord $\partial\Omega$. Afin de surmonter cette difficulté, généralement une information *a priori* sur la source supposée disponible. Dans le cadre de ce livre, nous nous intéressons à l'identification des sources ponctuelles dépendantes du temps c.-à-d. des sources de la forme suivante :

$$F(x,t) = \sum_{i=1}^{N} \lambda_i(t)\delta(x - S_i)$$

Avec $N \geq 1$ et pour $i = 1, .., N$, les S_i sont des points dans Ω qui représentent les positions des sources, $\lambda_i(.) \in L^2(0,T)$ désignent les fonctions de débits et δ représente la masse de Dirac.

Le dernier chapitre de ce livre discute le problème de la non-identifiabilité d'une source inconnue sous sa forme abstraite et propose une alternative qui consiste en l'identification de l'instant à partir duquel cette source inconnue a cessé d'émettre, sans aucune information *a priori* sur sa forme ni besoin de l'identifier complètement. En pratique, déterminer l'instant à partir duquel il n'y a plus de pollution introduite dans la rivière ne nécessite pas une information *a priori* sur la source impliquée ni à la déterminer complètement. Par ailleurs, cette détermination pourrait être un atout majeur et permettrait de lancer à temps des actions urgentes : commencer le nettoyage de la rivière, reprendre le pompage d'eau potable,...

Chapitre 2

Chapitre II

Identification of a point source in stationary $1D$ advection-dispersion-reaction equations with varying coefficients : detection of a pollution source

1 Introduction

One motivation for our study is an environmental application that consists of the identification of pollution sources in surface water : in a river, for example, the oxidaton of organic matter introduced by city sewages, industrial wastes,... usually drops to too low the level of the dissolved oxygen in the water. Since the lack of dissolved oxygen represents a serious threat to the diversity of the acquatic life, then localizing pollution sources and determining the intensity of the loaded organic matter could play a crucial role in preventing worse consequences regarding the perish of many acquatic species as well as in alerting downstream drinking water stations about the presence of an accidental pollution. This can be done by measuring the BOD (Biological Oxygen Demand) concentration which represents the amount of dissolved oxygen consumed by the microorganisms living in the river to decompose the introduced organic substances [4, 23]. Therefore, the more organic material there is, the higher the BOD concentration.

In the present study, we aim to localize the position and determine the intensity of a sought pollution source occurring in a portion of a river assimilated to a segment of a line using some measures of the BOD concentration. It is a nonlinear inverse source problem. Then, among the main questions concerning this study comes the following : How many measures of the BOD concentration do we need ? Where do those measures should be done in order to identify

Chapitre II. Identification of a point source in stationary $1D$ advection-dispersion-reaction equations with varying coefficients : detection of a pollution source

the elements defining the sought source in a unique manner? What about the stability of the identified results with respect to the perturbation of the used measures? The paper is organized as follows : Section 2 is devoted to stating the problem, assumptions and proving two technical lemmas for later use. In section 3, we prove the identifiability of the sought source from measuring the BOD concentration at two observation points framing the source region. Section 4 is reserved to establish an identification method that enables to localize the source position and determine its intensity using those two measures of the BOD concentration. In this section, we also prove the no identifiability of the elements defining the sought source if the two observation points do not frame the source region. Some numerical experiments on a variant of the surface water BOD pollution model are presented in section 5.

2 Mathematical modelling and problem statement

We consider a portion of a river assimilated to the segment $[0, \ell]$ and aim to identify a sought point source loading a pollution of constant intensity λ in a some location $S \in (0, \ell)$. The BOD concentration, denoted here by u, is governed by the following one-dimensional advection-dispersion-reaction equation, see [43, 50] :

$$-\Big(D(x)u'(x) - v(x)u(x)\Big)' + r(x)u(x) = \lambda \delta(x - S) \qquad \text{for} \quad 0 < x < \ell \qquad \text{(II.1)}$$

Where δ represents the Dirac mass, v is the flow velocity and D, r are respectively the diffusion and the reaction coefficients. Let $\mathcal{O}$ be an open interval of $\mathbb{R}$ containing $[0, \ell]$. Here, we assume r to be a continuous function on $\mathcal{O}$ and v to be a function of $\mathcal{C}^1$-class on $\mathcal{O}$ whereas D is assumed to be a positive and twice piecewise continuously differentiable function on $\mathcal{O}$. Then, the equation (II.1) is equivalent to

$$-D(x)u''(x) + V(x)u'(x) + R(x)u(x) = \lambda \delta(x - S) \qquad \text{for} \quad 0 < x < \ell \qquad \text{(II.2)}$$

With $V(x) := v(x) - D'(x)$ and $R(x) := r(x) + v'(x)$. As far as the boundary conditions are concerned and since the main transport is naturally oriented downstream, it seems to be reasonable the use of an homogeneous Dirichlet upstream boundary condition. However, at least two options are available for the downstream boundary condition : a null gradient concentration or simply a null concentration. This last option is usually employed when the downstream

II.2 Mathematical modelling and problem statement

boundary is assumed to be far away enough from the source position. In the present study, we use the first option namely $u'(\ell) = 0$. Therefore, in view of (II.2), the BOD concentration u satisfies the following system :

$$\begin{aligned} L[u](x) &= \lambda\delta(x-S) \quad \text{for} \quad 0 < x < \ell \\ u(0) &= u'(\ell) = 0 \end{aligned} \quad\quad\quad (\text{II.3})$$

Where L is the second order linear differential operator defined as follows :

$$L[u](x) := -D(x)u''(x) + V(x)u'(x) + R(x)u(x) \quad\quad\quad (\text{II.4})$$

Notice that due to the linearity of the operator L introduced in (II.4) and in view of the superposition principle, the use of inhomogeneous boundary conditions do not affect the results established in this paper.

Besides, it is well known that with the used regularity conditions on the coefficients D, v and r, the problem (II.3) admits a unique solution $u \in H^1(0,\ell)$, see [47, 57]. Therefore, u is a continuous function and we can use its value at any point x of $(0,\ell)$. Then, given two points a and b such that $0 < a < b < \ell$, we define the following observation operator :

$$M[S,\lambda] := \{u(a), u(b)\} \quad\quad\quad (\text{II.5})$$

This is the so-called *direct problem*. The *inverse problem* with which we are concerned here is : assuming available the data $\{d_a, d_b\}$, find the source parameters S and λ such that

$$M[S,\lambda] = \{d_a, d_b\} \quad\quad\quad (\text{II.6})$$

In the literature, this inverse source problem in one-dimensional evolution advection-dispersion-reaction equations with constant diffusion, reaction and velocity coefficients was treated in [17, 30] for a single model and in [19, 31] for a coupled model. In addition, the two-dimensional single model with constant coefficients was studied in [29] for the stationary case and recently in [33] for the evolution case. Furthermore, similar inverse source problems for the heat equation are treated in [2, 3, 12, 16, 38, 44, 47]. The originality of the present study consists in studying

Chapitre II. Identification of a point source in stationary 1D advection-dispersion-reaction equations with varying coefficients : detection of a pollution source

the underlined inverse source problem in transport equations with varying diffusion, reaction and velocity coefficients.

For reasons to be explained later, let us introduce the function u_0 solution to the following system :

$$
\begin{aligned}
L[u_0](x) = 0 \quad &\text{for } x \in \mathcal{O} \\
u_0(x_0) = 0 \quad &\text{and} \quad u_0(\ell) = 1
\end{aligned}
\qquad (\text{II.7})
$$

Where $x_0 < 0$ and $x_0 \in \mathcal{O}$. In addition, since $D(x) > 0$ for all $x \in \mathcal{O}$, we also introduce the function

$$
p(x) = e^{-\int_0^x \frac{V(\eta)}{D(\eta)} d\eta} \quad \text{for} \quad x \in \mathcal{O} \qquad (\text{II.8})
$$

Then, we establish the following lemma that states the main condition on the varying flow characteristics D, v and r needed to prove the identifiability of the sought point source :

Lemma 2.1 *If the varying coefficients D, v and r satisfy the following condition :*

$$
4r(x) + \frac{v^2(x)}{D(x)} + 2v'(x) + 2D''(x) \geq \frac{(D'(x))^2}{D(x)} \quad \text{for all} \quad x \in \mathcal{O} \qquad (\text{II.9})
$$

Then, the function u_0 solution to the system (II.7) doesn't admit any root in $(0, \ell)$.

Proof. Since $D(x) > 0$ for all $x \in \mathcal{O}$ and with refer to (II.4), it comes that the function u_0 solution to (II.7) satisfies

$$
u_0''(x) - \frac{V(x)}{D(x)} u_0'(x) - \frac{R(x)}{D(x)} u_0(x) = 0 \quad \text{for all} \quad x \in \mathcal{O} \qquad (\text{II.10})
$$

Using the function p introduced in (II.8) and the change of variable : $\tilde{u}_0(x) = \sqrt{p(x)} u_0(x)$ for $x \in \mathcal{O}$, the equation (II.10) is equivalent to

$$
\tilde{u}_0''(x) + g(x)\tilde{u}_0(x) = 0 \quad \text{for} \quad x \in \mathcal{O} \qquad (\text{II.11})
$$

Where g is the function defined for all x in $\mathcal{O}$ as follows :

II.2 Mathematical modelling and problem statement

$$\begin{aligned}
g(x) &= -\frac{R(x)}{D(x)} - \frac{1}{2}\frac{p''(x)}{p(x)} + \frac{1}{4}\left(\frac{p'(x)}{p(x)}\right)^2 \\
&= -\frac{R(x)}{D(x)} - \frac{1}{4}\left(\frac{V(x)}{D(x)}\right)^2 + \frac{1}{2}\left(\frac{V(x)}{D(x)}\right)' \\
&= -\frac{1}{4D(x)}\left(4r(x) + \frac{v^2(x)}{D(x)} + 2v'(x) + 2D''(x) - \frac{(D'(x))^2}{D(x)}\right)
\end{aligned} \qquad (\text{II}.12)$$

In view of the last equality in (II.12), the assertion (II.9) yields $g(x) \leq 0$ for all $x \in \mathcal{O}$. Therefore, as proved in [11], all solution $\tilde{u}_0$ to (II.11) is a nonoscillating function in $\mathcal{O}$. That means $\tilde{u}_0$ has at most one root in the open interval $\mathcal{O}$. Furthermore, as u_0 and $\tilde{u}_0$ have the same roots and $u_0(x_0) = 0$, we conclude that $u_0(x) \neq 0$ for all x in $(0, \ell)$. $\square$

Then, using the function u_0 solution to the system (II.7), we express the state u solution to the problem (II.3) as follows :

Lemma 2.2 *Provided Lemma 2.1 applies, the function u solution to the problem (II.3) is defined by*

$$u(x) = \frac{\lambda p(S) u_0(S)}{D(S)} u_0(x) \left(\left[1 - \beta \psi(S)\right] \psi(x) - \mathcal{H}(x-S)\left[\psi(x) - \psi(S)\right] \right) \qquad (\text{II}.13)$$

With $\mathcal{H}$ is the Heaviside function, p is the function introduced in (II.8) and u_0 is the solution to the system introduced in (II.7) whereas the function ψ and the constant β are as follows :

$$\psi(x) = \int_0^x \frac{1}{p(\eta) u_0^2(\eta)} d\eta \quad \text{and} \quad \beta = \frac{u_0'(\ell) p(\ell)}{1 + u_0'(\ell) p(\ell) \psi(\ell)} \qquad (\text{II}.14)$$

Proof. Let u_0 be the solution to (II.7) and Φ be the function such that : $u(x) = \Phi(x) u_0(x)$. The derivative function $\varphi = \Phi'$ satisfies in $(0, \ell)$ the following equation :

$$-Du_0 \varphi' + \left(Vu_0 - 2Du_0'\right)\varphi = \lambda \delta(x - S) \qquad (\text{II}.15)$$

Furthermore, using the function p introduced in (II.8) and u_0 we find that the function

Chapitre II. Identification of a point source in stationary 1D advection-dispersion-reaction equations with varying coefficients : detection of a pollution source

$$\varphi_0(x) = \frac{1}{p(x)u_0^2(x)} \quad \text{solves} \quad -Du_0\varphi_0' + \left(Vu_0 - 2Du_0'\right)\varphi_0 = 0 \tag{II.16}$$

Therefore, the function φ defined for all constant c as follows :

$$\varphi(x) = \varphi_0(x)\left(c - \frac{\lambda p(S)u_0(S)}{D(S)}\mathcal{H}(x-S)\right) \tag{II.17}$$

Solves the equation introduced in (II.15). Notice that in (II.17), $\mathcal{H}$ represents the Heaviside function. As the state u satisfies $u(0) = 0$, we set $u(x) = u_0(x)\Phi(x) = u_0(x)\int_0^x \varphi(\eta)d\eta$. Then, from (II.17) we obtain

$$u(x) = u_0(x)\left(c\int_0^x \varphi_0(\eta)d\eta - \frac{\lambda p(S)u_0(S)}{D(S)}\mathcal{H}(x-S)\int_S^x \varphi_0(\eta)d\eta\right) \tag{II.18}$$

In addition, we have $u'(\ell) = u_0'(\ell)\Phi(\ell) + u_0(\ell)\varphi(\ell)$ and according to (II.7) we have also $u_0(\ell) = 1$. Hence, to satisfy the downstream boundary condition $u'(\ell) = 0$ we select the constant c involved in (II.17)-(II.18) such that $\varphi(\ell) + \Phi(\ell)u_0'(\ell) = 0$. That implies, we set

$$c = \frac{\lambda p(S)u_0(S)}{D(S)}\left(1 - \beta\psi(S)\right) \tag{II.19}$$

Where β and ψ are as introduced in (II.14). And thus, from (II.18) and (II.19) we find the results announced in (II.13)-(II.14).

□

3 Identifiability

Provided the varying coefficients D, v and r satisfy the main condition (II.9) and the two observation points a, b frame the source region, we prove in this section that the elements S and λ defining the sought point source are determined in a unique manner using the observation operator $M[S, \lambda]$ introduced in (II.5).

Theorem 3.1 *Let $0 < a < b < \ell$ and for $i = 1, 2$, $M[S_i, \lambda_i]$ be the observation operator*

II.3 Identifiability

introduced in (II.5) and associated to the state u_i solution to the problem (II.3) with the point source $\lambda_i \delta(x - S_i)$. If the condition (II.9) holds true and for $i = 1, 2$ we have $a < S_i < b$ then,

$$M[S_1, \lambda_1] = M[S_2, \lambda_2] \implies S_1 = S_2 \text{ and } \lambda_1 = \lambda_2 \qquad (\text{II.20})$$

Proof. For $i = 1, 2$, let u_i be the solution to the problem introduced in (II.3) with the point source $\lambda_i \delta(x - S_i)$. The variable $w = u_2 - u_1$ satisfies the following system :

$$\begin{aligned} L[w](x) &= \lambda_2 \delta(x - S_2) - \lambda_1 \delta(x - S_1) \quad \text{for } 0 < x < \ell \\ w(0) &= w'(\ell) = 0 \end{aligned} \qquad (\text{II.21})$$

Then, using the same notations and techniques as employed to prove Lemma 4.1 we determine the solution w to the system (II.21) as follows :

$$w(x) = u_0(x) \sum_{i=1}^{2} (-1)^i \alpha_i \Big([1 - \beta \psi(S_i)] \psi(x) - \mathcal{H}(x - S_i)[\psi(x) - \psi(S_i)] \Big) \text{ where } \alpha_i = \frac{\lambda_i p(S_i) u_0(S_i)}{D(S_i)} \qquad (\text{II.22})$$

Moreover, in view of (II.22) and since we have $0 < a < S_i < b < \ell$ for $i = 1, 2$ we obtain

$$\begin{aligned} w(a) &= u_0(a) \psi(a) \Big(\alpha_2 \big[1 - \beta \psi(S_2)\big] - \alpha_1 \big[1 - \beta \psi(S_1)\big] \Big) \\ w(b) &= u_0(b) \big(1 - \beta \psi(b)\big) \Big(\alpha_2 \psi(S_2) - \alpha_1 \psi(S_1) \Big) \end{aligned} \qquad (\text{II.23})$$

Besides, according to the observation operator introduced in (II.5), we have $M[S_2, \lambda_2] = M[S_1, \lambda_1]$ implies that $u_2(a) = u_1(a)$ and $u_2(b) = u_1(b)$ which leads to $w(a) = w(b) = 0$. Furthermore, since with refer to Lemma 2.1 we have $u_0(x) \neq 0$ for all $x \in (0, \ell)$ and in view of (II.14) we have $1 - \beta \psi(b) \neq 0$, the second equation in (II.23) gives $\alpha_2 \psi(S_2) = \alpha_1 \psi(S_1)$. Then, using this last result in the first equation of (II.23) we find $\alpha_1 = \alpha_2$. That implies, we have $\psi(S_2) = \psi(S_1)$. Therefore, as the function ψ introduced in (II.14) is continuous and strictly increasing in (a, b), we obtain $S_1 = S_2$. In addition, using this last result in the value of α_i introduced in (II.22) and since $\alpha_1 = \alpha_2$ we get $\lambda_1 = \lambda_2$.

Chapitre II. Identification of a point source in stationary $1D$ **advection-dispersion-reaction equations with varying coefficients : detection of a pollution source**

4 Identification

In this section, we establish an identification method that enables to determine the elements S and λ defining the sought point source from measuring the state u at two observation points framing the source region. However, if the two observation points do not frame the source region (they are both situated upstream or downstream with respect to the source location), we prove that the measures of u taken at those two points do not enable to identify the elements S and λ in a unique manner.

Theorem 4.1 *Let u be the solution to the problem (II.3) with a point source $\lambda\delta(x-S)$ and $M[S,\lambda]$ be the observation operator introduced in (II.5) using the two points a, b. Provided Lemma 2.1 applies and $0 < a < S < b < \ell$, if $M[S,\lambda] = \{d_a, d_b\}$ then S and λ are subject to :*

$$\psi(S) = Q \quad \text{and} \quad \lambda = \frac{d_b D(S)}{p(S)u_0(S)u_0(b)Q\big(1-\beta\psi(b)\big)} \tag{II.24}$$

With p is the function introduced in (II.8) and ψ, β are given in (II.14) whereas the constant Q is such that

$$Q = \frac{d_b u_0(a)\psi(a)}{d_a u_0(b) + \beta\big(d_b u_0(a)\psi(a) - d_a u_0(b)\psi(b)\big)} \tag{II.25}$$

Proof. Let u be the solution to the problem (II.3) with a point source $\lambda\delta(x-S)$ and a, b be two observation points of $(0,\ell)$ such that $0 < a < S < b < \ell$. Since in view of (II.5) we have $M[S,\lambda] = \{d_a, d_b\}$ implies that $u(a) = d_a$ and $u(b) = d_b$, then according to the results (II.13)-(II.14) of Lemma 4.1 we obtain

$$\begin{aligned} d_a &= \lambda \frac{p(S)u_0(S)}{D(S)} u_0(a)\big(1-\beta\psi(S)\big)\psi(a) \\ d_b &= \lambda \frac{p(S)u_0(S)}{D(S)} u_0(b)\big(1-\beta\psi(b)\big)\psi(S) \end{aligned} \tag{II.26}$$

Moreover, from dividing the second by the first equation in (II.26) we find

II.4 Identification

$$\frac{d_b}{d_a} = \frac{u_0(b)\left(1 - \beta\psi(b)\right)\psi(S)}{u_0(a)\left(1 - \beta\psi(S)\right)\psi(a)} \tag{II.27}$$

Therefore, using the equation (II.27) we obtain the first result announced in (II.24). In addition, from the second equation of the system (II.26) we get the source intensity λ introduced in (II.24). □

In the remaining part of this section, we prove the non identifiability of the elements S and λ defining the sought source if the two observation points a and b do not frame the source region. This result is given by the following corollary :

Corollary 4.2 *If the two observation points a and b do not frame the source region then, the observation operator $M[S, \lambda]$ introduced in (II.5) does not enable to identify the two unknowns S and λ in a unique manner.*

Proof. Let u be the solution to the problem (II.3) with a point source $\lambda\delta(x-S)$ and $M[S, \lambda]$ be the observation operator introduced in (II.5) using the points a and b. If those two points are such that $0 < a < b < S < \ell$ then, having $M[S, \lambda] = \{d_a, d_b\}$ implies that $u(a) = d_a$ and $u(b) = d_b$ which according to (II.13)-(II.14) of Lemma 4.1 leads to the following system of two linearly dependent equations :

$$\begin{aligned} d_a &= \lambda \frac{p(S)u_0(S)}{D(S)}\left(1 - \beta\psi(S)\right)u_0(a)\psi(a) \\ d_b &= \lambda \frac{p(S)u_0(S)}{D(S)}\left(1 - \beta\psi(S)\right)u_0(b)\psi(b) \end{aligned} \tag{II.28}$$

Besides, if the two observation points a and b are such that $0 < S < a < b < \ell$ then, $M[S, \lambda] = \{d_a, d_b\}$ gives $u(a) = d_a$, $u(b) = d_b$ and using (II.13)-(II.14) we obtain

$$\begin{aligned} d_a &= \lambda \frac{p(S)u_0(S)}{D(S)}\psi(S)u_0(a)\left(1 - \beta\psi(a)\right) \\ d_b &= \lambda \frac{p(S)u_0(S)}{D(S)}\psi(S)u_0(b)\left(1 - \beta\psi(b)\right) \end{aligned} \tag{II.29}$$

The system (II.29) is also a system of two linearly dependent equations. Therefore, in both cases where the two observation points a and b do not frame the source region, having

Chapitre II. Identification of a point source in stationary 1D advection-dispersion-reaction equations with varying coefficients : detection of a pollution source

$M[S, \lambda] = \{d_a, d_b\}$ leads to a system of two linearly dependent equations which is not enough to determine the two unknowns S and λ in a unique manner. □

5 Numerical experiments

We start this section by deriving the undimensioned version of the identification problem. To this end, we employ the variable $y = x/\ell$. Then, using the coefficients $\hat{D}(y) = D(x)$, $\hat{v}(y) = v(x)$ and $\hat{r}(y) = r(x)$, the BOD concentration $\hat{u}(y) = u(x)$ reduced to the interval $[0, 1]$ satisfies the following system :

$$\begin{array}{ll} -\hat{D}(y)\hat{u}''(y) + \hat{V}(y)\hat{u}'(y) + \hat{R}(y)\hat{u}(y) = \ell^2 \lambda \delta(x - \hat{S}) & \text{for} \quad 0 < y < 1 \\ \hat{u}(0) = \hat{u}'(1) = 0 \end{array} \tag{II.30}$$

With $\hat{V}(y) = \ell\hat{v}(y) - \hat{D}'(y)$ and $\hat{R}(y) = \ell\big(\ell\hat{r}(y) + \hat{v}'(y)\big)$ whereas $\hat{S} = S/\ell$. Then, according to Theorem 4.1 and given the measures $\{\hat{d}_{\hat{a}}, \hat{d}_{\hat{b}}\}$ of $\hat{u}$ taken at the two observation points $\hat{a} = a/\ell$ and $\hat{b} = b/\ell$ such that $0 < \hat{a} < \hat{S} < \hat{b} < 1$, the source parameters $\hat{S}$ and λ are subject to :

$$\hat{\psi}(\hat{S}) = \hat{Q} \quad \text{and} \quad \lambda = \frac{\hat{d}_{\hat{b}}\hat{D}(\hat{S})}{\ell^2 \hat{p}(\hat{S})\hat{u}_0(\hat{S})\hat{u}_0(\hat{b})\hat{Q}\big(1 - \hat{\beta}\hat{\psi}(\hat{b})\big)} \tag{II.31}$$

With $\hat{p}(y) = e^{-\int_0^y \frac{\hat{V}(\xi)}{\hat{D}(\xi)}d\xi}$ and $\hat{\beta} = \hat{u}'_0(1)\hat{p}(1)/\big(1 + \hat{u}'_0(1)\hat{p}(1)\hat{\psi}(1)\big)$ whereas $\hat{Q}$ and the function $\hat{\psi}$ are such that

$$\hat{\psi}(y) := \int_0^y \frac{1}{\hat{p}(\xi)\hat{u}_0^2(\xi)} d\xi \quad \text{and} \quad \hat{Q} = \frac{\hat{d}_{\hat{b}}\hat{u}_0(\hat{a})\hat{\psi}(\hat{a})}{\hat{d}_{\hat{a}}\hat{u}_0(\hat{b}) + \hat{\beta}\big(\hat{d}_{\hat{b}}\hat{u}_0(\hat{a})\hat{\psi}(\hat{a}) - \hat{d}_{\hat{a}}\hat{u}_0(\hat{b})\hat{\psi}(\hat{b})\big)} \tag{II.32}$$

Here we have the function $\hat{u}_0(y) = u_0(x)$ where u_0 is the solution to the problem (II.7). To compute $\hat{u}_0(y)$ and using similar techniques as employed in the proof of Lemma 2.1 namely using the change of variable $z(y) = \sqrt{\hat{p}(y)}\hat{u}_0(y)$, we solve the following system :

$$\begin{array}{l} z''(y) + \hat{g}(y)z(y) = 0 \\ z(\hat{x}_0) = 0 \quad \text{and} \quad z(1) = \sqrt{\hat{p}(1)} \end{array} \tag{II.33}$$

II.5 Numerical experiments

Where $\hat{x}_0 = x_0/\ell$ and the function $\hat{g}$ is defined as follows :

$$\hat{g}(y) = -\frac{1}{4\hat{D}(y)}\left(4\ell^2 \hat{r}(y) + \ell^2 \frac{\hat{v}^2(y)}{\hat{D}(y)} + 2\ell\hat{v}'(y) + 2\hat{D}''(y) - \frac{\left(\hat{D}'(y)\right)^2}{\hat{D}(y)}\right) \quad (II.34)$$

Given $N > 0$, we use a uniform discretization of the segment $[0,1]$ by considering the discrete points : $y_i = i*h$ for $i = 0, ..., N+1$ where the step size $h = 1/(N+1)$. Then, we set $\hat{x}_0 = -h$ and by employing the three-points finite differences method, we obtain from (II.33) the following linear system :

$$A(z_0, ..., z_N)^\top = \left(0, 0, ..., -\sqrt{\hat{p}(1)}\right)^\top \quad (II.35)$$

Where $z_i \approx z(y_i)$ for $i = 0, ..., N$ and $A = tridiag\left(1, h^2 \hat{g}(y_i) - 2, 1\right)$. Furthermore, we determine the unknown vector $(z_0, ..., z_N)$ solution to (II.35) using the Gauss-Seidel method. For numerical experiments, we use the following diffusion function, see [51] : $D(x) = d_m + \theta\left(1 - e^{-\gamma x}\right)$ where d_m is the molecular diffusion and θ, γ are two given positive real numbers. Besides, we employ a mean velocity value v_0 and a constant reaction coefficient r_0. Therefore, the function $\hat{p}(y) = p(x)$ associated to the function p introduced in (II.8) is given by

$$\hat{p}(y) = e^{-\int_0^y \frac{\hat{v}(\xi)}{\hat{D}(\xi)}d\xi} = \left(\frac{d_m}{(d_m + \alpha)e^{\gamma\ell y} - \theta}\right)^{\left(\frac{v_0}{\beta(d_m + \theta)} - 1\right)} e^{-\gamma\ell y} \quad (II.36)$$

The second equality in (II.36) is obtained using the change of variable : $\zeta = e^{\gamma\ell\xi}$. In addition, referring to (II.34) and in order to satisfy the main condition (II.9) namely $\hat{g} \leq 0$ in an open interval containing $[0,1]$, we introduce the function Γ defined from $\hat{g}$ as follows :

$$\begin{aligned}\Gamma(y) &= -4\hat{D}(y)^2 \hat{g}(y) \\ &= 4\ell^2 r_0 \hat{D}(y) + \ell^2 v_0^2 + 2\hat{D}(y)\hat{D}''(y) - \left(\hat{D}'(y)\right)^2 \\ &= \ell^2\left(4r_0(d_m + \theta) + v_0^2 - 2\theta\left(2r_0 + \gamma^2(d_m + \theta)\right)e^{-\gamma\ell y} + \theta^2\gamma^2 e^{-2\gamma\ell y}\right)\end{aligned} \quad (II.37)$$

Then, using the last equality in (II.37), it comes that the function Γ has exactly one extrumum in $I\!R$ which is a minimum reached at the point :

Chapitre II. Identification of a point source in stationary 1D advection-dispersion-reaction equations with varying coefficients : detection of a pollution source

$$y^* = -\frac{1}{\gamma \ell} \ln\left(\frac{2r_0 + \gamma^2(d_m + \theta)}{\theta \gamma^2}\right) \quad \text{with} \quad \Gamma(y^*) = v_0^2 - \left(4\left(\frac{r_0}{\gamma}\right)^2 + \gamma^2(d_m + \theta)^2\right) \quad \text{(II.38)}$$

Hence, as the two functions Γ and $\hat{g}$ are of opposite signs, we obtain

$$v_0 \geq \sqrt{4\left(\frac{r_0}{\gamma}\right)^2 + \gamma^2(d_m + \theta)^2} \quad \Longrightarrow \quad \hat{g}(y) \leq 0 \quad \text{for all } y \quad \text{(II.39)}$$

Thus, provided (II.39) holds true and according to Lemma 2.1, the function z solution to the problem (II.33) does not admit any root in $(0,1)$ which implies that the function $\hat{u}_0 = \hat{p}^{-1/2} z$ also does not admit any root in $(0,1)$. Therefore, in view of (II.31)-(II.32), we determine the source position $\hat{S}$ as the zero in $(\hat{a}, \hat{b})$ of the function $y \mapsto \int_0^y \left(\hat{p}(\xi)\hat{u}_0^2(\xi)\right)^{-1} d\xi - \hat{Q}$. To this end, we use the Newton method. Once $\hat{S}$ is known, we determine the sought intensity λ from (II.31).

To carry out numerical experiments, we use $\ell = 1000m$, the diffusion parameters : $d_m = 10^{-5}m^2s^{-1}$, $\theta = 10$, $\gamma = 9 \times 10^{-3}$, the mean velocity $v_0 = 0.1ms^{-1}$ and the reaction coefficient $r_0 = 10^{-5}s^{-1}$. We discretize the interval $[0,1]$ using a uniform step size $h = 1/100$. Then, to give a numerical explanation to the non identifiability of the elements defining the sought point source when the two observation points do not frame the source region, we present the behaviour of the BOD concentration obtained from the results (II.13)-(II.14) of Lemma 4.1 with $S = 563m$ and $\lambda = 2.41 \times 10^{-5}gl^{-1}s^{-1}$. This behaviour is given by the following graph :

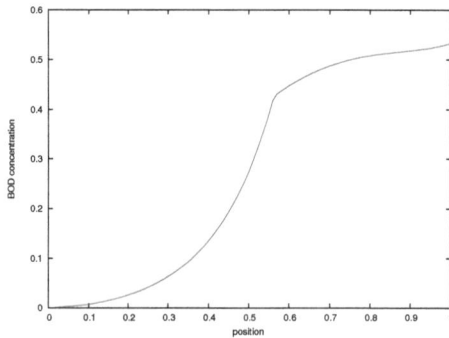

Figure II.1 – BOD concentration : $S = 563\ m$ and $\lambda = 2.41 \times 10^{-5}gl^{-1}s^{-1}$

II.5 Numerical experiments

The behaviour of the BOD concentration presented in Figure II.1 reveals that having the two observation points situated both in a same side of the source location will enable to detecte only a small variation of the BOD concentration but certainly not its significant variation which characterises the sought source. However, this significant variation is well detected if the two observation points frame the source region. This analysis is confirmed by the identifiability result introduced in Theorem 3.1 and the no identifiability result established in Corollary 4.2

In the remainder of this section, we employ the two observation points $a = 200\ m, b = 800\ m$ and aim to study the stability of the established identification method with respect to the introduction of a perturbation on the used measures. To this end, we study the behaviour of the relative errors on the identified source position $|S - S_{ident}|/S$ and on the identified source intensity $|\lambda - \lambda_{ident}|/\lambda$ with respect to the introduction of a Gaussian noise on the used measures d_a and d_b. Then, we introduced different intensities of a Gaussian noise on the measures obtained from solving the problem (II.3)-(II.4) with the source parameters : $S = 647\ m$ and $\lambda = 3.56 \times 10^{-5} gl^{-1}s^{-1}$ and compute in each time the identified source parameters S_{ident} and λ_{ident} using the established identification results introduced in (II.24)-(II.25). The results of this study are presented in the following graph :

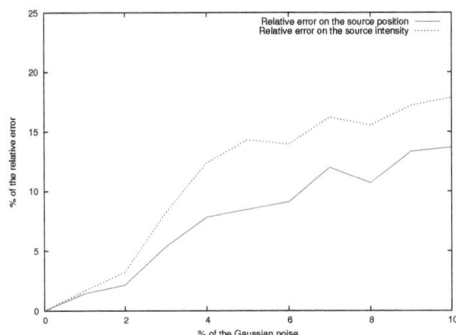

Figure II.2 – Relative errors on the identified source parameters

The analysis of the results presented in Figure II.2 shows that the established identification method enables to localize the sought source and determine its intensity with a good accuracy. Furthermore, from the graph presented in Figure II.2 we observe that the relative error on the identified source intensity seems to be bigger than the relative error on the identified source position. This observation could be explained by the fact that we use the identified value S_{ident}

Chapitre II. Identification of a point source in stationary $1D$ advection-dispersion-reaction equations with varying coefficients : detection of a pollution source

of the source position in order to determine the intensity lambda λ_{ident} from the second result in (II.24). In practice and for an accidental pollution, usually we have to identify an active source among some well known suspected locations. Then, having S_{ident} will likely lead to determine the exact location S and thus to improve the relative error on the identified source intensity by using the exact value S in the determination of λ_{ident}.

6   Conclusion

In this paper, we studied the identification of a point source occurring in the right handside of a one-dimensional advection-dispersion-reaction equation with varying coefficients. We derived a main condition on those varying coefficients that implies the identifiability of the elements defining the sought source from measuring the state at two observation points framing the source region. We also proved the non identifiability of those elements if the two observation points are both situated in a same side of the source location (both upstream or both downstream). We established and identification method that enables to localize the sought source and determine its intensity. We carried out some numerical experiments on a variant of the surface water BOD pollution model. The analysis of those experiments shows that the established identification method identifies the sought point source with a good accuracy and is stable with respect to the introduction of a Gaussian noise on the used measures.

Chapitre III

Inverse source problem in a one-dimensional evolution linear transport equation with spatially varying coefficients : application to surface water pollution

Les résultats de ce chapitre sont issus de l'article [35], publié en collaboration avec Adel Hamdi dans la revue *Inverse Problems in Science and Engineering*.

1 Introduction

Inverse problems play a key role in providing estimations of unknown and sometimes even inaccessible elements involved in the associated mathematical model using some observations of its response. In real world problems, having an accurate estimation of the missing elements in the mathematical model usually leads to a better understanding of the occurring phenomena and thus, to take appropriate actions in order to prevent undesirable situations. During the last few decades, we have seen inverse problems to be employed in numerous areas of science and engineering : in medicine, the inverse problem of electrocardiography, for example, is used to restore the heart activity from a given set of body surface potentials [64]. In seismology, inverse source problems are used to determine the hypocenter of an earthquake [40] as well as to study the dynamic problem of seismology which is one of the most topical problems of geophysical [6].

A motivation for our present study concerning inverse source problems in transport equations is a typical problem associated with environmental monitoring which can be described as follows : certain areas like water, groundwater or atmosphere can be monitored by some sensors

Chapitre III. Inverse source problem in a one-dimensional evolution linear transport equation with spatially varying coefficients : application to surface water pollution

destinated to evaluate the level of pollution in the site. When the incoming signals reveal an unusual rise in pollution concentrations, the top priority action becomes the identification of the contamination source as quickly as possible in order to prevent worse consequences. A concrete example of this situation consists of the identification of pollution sources in surface water : in a river, for example, the oxidaton of organic matter introduced by city sewages, industrial wastes,... usually drops to too low the level of dissolved oxygen DO in the water. Problems with low concentrations of DO are essentially an unbalanced ecosystem with fish mortality, odours and other nuisances, see [13] for more details. Therefore, as soon as the sensors begin to inform about a lack of DO, the identification of pollution sources becomes a priority in order to preserve the diversity of the acquatic life and prevent the perish of many species. That enables also to alert downstream drinking water stations about the presence of an accidental pollution. The identification of sought pollution sources in a river could be done by monitoring the *Biological Oxygen Demand* (*BOD*) concentration which represents the amount of dissolved oxygen consumed by the micro-organisms living in the river to decompose the introduced organic substances [4, 23]. Thus, the more organic material there is, the higher the BOD concentration.

In this paper, we assume monitoring a portion of a river assimilated to a segment of a line and are interested in the identification of an unknown pollution source responsable of the higher BOD concentrations recorded by some sensors already placed in this portion. The paper is organized as follows : section 2 is devoted to stating the problem, assumptions and proving some technical results for later use. In section 3, we prove under some reasonable assumptions the identifiability of the sought source from recording the generated state at two observation points framing the source region. Section 4 is reserved to establish an identification method that uses those records to determine the elements defining the sought source. Some numerical experiments on a variant of the surface water BOD pollution model are presented in section 5.

2 Mathematical modelling and problem statement

We suppose monitoring a portion of a river represented by the segment $(0, l)$ during a time $T > 0$. The BOD concentration, denoted here by u, in this portion is governed by the following one-dimensional parabolic partial differential equation, see [43, 50] :

$$\partial_t u(x,t) - \partial_x \Big(D(x) \partial_x u(x,t) - v(x) u(x,t) \Big) + r u(x,t) = F(x,t) \qquad \text{(III.1)}$$

III.2 Mathematical modelling and problem statement

Where F represents the pollution source, v is the flow velocity and D, r are respectively the dispersion and reaction coefficients. Here, r is a positive real number, v is a function of $\mathcal{C}^1$-class on $[0, l]$ whereas D is a twice piecewise continuously differentiable function on $[0, l]$ such that $D(x) > 0$ for all $x \in [0, l]$. The equation (III.1) is equivalent to

$$\partial_t u(x,t) - D(x)\partial_{xx}u(x,t) + V(x)\partial_x u(x,t) + R(x)u(x,t) = F(x,t) \tag{III.2}$$

With $V(x) = v(x) - D'(x)$ and $R(x) = r + v'(x)$. Then, multiplying (III.2) by the weight function ℓ defined as follows:

$$\ell(x) = \frac{p(x)}{D(x)} \quad \text{where} \quad p(x) = e^{-\int_0^x \frac{V(\eta)}{D(\eta)} d\eta} \tag{III.3}$$

Implies that the BOD concentration u satisfies

$$L[u](x,t) = \ell(x) F(x,t) \quad \text{for} \quad 0 < x < l, \quad 0 < t < T \tag{III.4}$$

Where L is the following parabolic differential operator:

$$L[u](x,t) = \ell(x)\partial_t u(x,t) - \partial_x\Big(p(x)\partial_x u(x,t)\Big) + q(x)u(x,t) \tag{III.5}$$

With $q(x) = \ell(x)R(x)$. As far as initial and boundary conditions are concerned, one could consider without loss of generality no pollution occurring at the initial monitoring time and thus, a null initial BOD concentration. In addition, as the main transport is naturally oriented downstream, it seems to be reasonable the use of an homogeneous Dirichlet upstream boundary condition. However, at least two options are available for the downstream boundary condition: a null gradient concentration or simply a null concentration. This last option is usually employed when the downstream boundary is assumed to be far away enough from the source position. In this paper, we use the following homogeneous initial and boundary conditions:

$$u(x,0) = 0 \quad \text{for} \quad 0 < x < l \tag{III.6}$$

$$u(0,t) = 0 \quad \text{and} \quad \partial_x u(l,t) = 0 \quad \text{for} \quad 0 < t < T \tag{III.7}$$

Chapitre III. Inverse source problem in a one-dimensional evolution linear transport equation with spatially varying coefficients : application to surface water pollution

Notice that due to the linearity of the operator L introduced in (III.5) and in view of the superposition principle, the use of a non-zero initial condition and/or inhomogeneous boundary conditions do not affect the results established in this paper.

Furthermore, it is well known that under reasonable assumptions on the regularity of the source F, the problem (III.4)-(III.7) admits a unique solution u smooth enough to use its value at any point (x,t) of $(0,l) \times (0,T)$, see [47]. Therefore, given two observation points a and b such that $0 < a < b < l$, we can define the following observation operator :

$$M[F] := \{u(a,t),\ u(b,t)\ \text{ for }\ 0 < t < T\} \tag{III.8}$$

This is the so-called *direct problem*.

The *inverse problem* with which we are concerned here is : assuming available the records $\{d_a(t),\ d_b(t)\ \text{ for }\ 0 < t < T\}$ of the concentration u at the two observation points a and b, find the source F such that

$$M[F] = \{d_a(t),\ d_b(t)\ \text{ for }\ 0 < t < T\} \tag{III.9}$$

The main difficulty in such kind of inverse problem is that in general there is no identifiability of the source F in its abstract form, see [17]. In the literature, to overcome this difficulty authors generally assume available some *a priori* information on the source F : For example, time-independent sources $F(x,t) = f(x)$ are treated by J.R. Cannon in [12] using spectral theory, then by H. Engl, O. Scherzer and M. Yamamoto in [21] using the approximated controllability of the heat equation. The results of this last paper are generalized by M. Yamamoto in [65, 66] to sources of the form $F(x,t) = \alpha(t)f(x)$ where $f \in L^2$ and the time-dependent function $\alpha \in C^1[0,T]$ is assumed to be known and satisfying the condition $\alpha(0) \neq 0$. Furthermore, F. Hettlich and W. Rundell addressed in [38] the $2D$ inverse source problem for the heat equation with sources of the form $F(x,t) = \chi_\mathcal{D}(x)$ where $\mathcal{D}$ is a subset of a disk. They proved the identifiability of $\mathcal{D}$ from recording the flux at two different points of the boundary. A. El Badia and A. Hamdi studied in [17, 19] for a one-dimensional evolution linear transport equation with constant diffusion, velocity and reaction coefficients the identification of a time-dependent point source $F(x,t) = \lambda(t)\delta(x - S)$ where the source position S and the time-dependent intensity function λ are both unknown. They proved the identifiability of F from recording the state

III.2 Mathematical modelling and problem statement

and its flux at two observation points framing the source region. Those results for the case of linear transport equations with constant coefficients have been recently improved by A. Hamdi in [30, 31] to requiring only the record of the state at the two observation points.

The originality of the present study with respect to [30, 31] consists in considering the underlined inverse source problem in the general case of linear evolution transport equations with spatially varying diffusion, velocity and reaction coefficients. That increases the degree of difficulty and makes the results established in [30, 31] with constant coefficients do not apply at least for the two following reasons : **1.** In [30, 31], the essential ingredient of localizing quasi-explicitly the position of the sought source is the use of the impulse response to the operator that is the adjoint of the spatial part of the operator introduced in the left-hand side of (III.2) with constant D, V and R coefficients. Then, this impulse response is explicitly determined as the solution to a second order linear differential equation with constant coefficients. In the present study, that does not apply with arbitrary spatially varying D, V and R coefficients **2.** In [30, 31], by employing a change of variable, the operator in the left-hand side of (III.2) with constant D, V and R coefficients is transformed into a symmetric operator (the heat equation). And thus, to recover the source intensity function, one solves a deconvolution problem where the associated state is expressed in the complete orthogonal family made by the classic Laplacian eigenfunctions. In this paper, the nonsymmetry in the spatial part of the operator introduced in the left-hand side of (III.2) requires the determination of an adequate weight function that transforms the problem of finding a complete orthogonal family into solving a generalized Sturm-Liouville eigenvalue problem. Then, conditions on the spatially varying coefficients need to be found in order to deal with a regular Sturm-Liouville problem.

According to the usual mathematical modelling of a time-dependent point source, we use in this paper a source F that takes the form

$$F(x,t) = \lambda(t)\,\delta(x-S) \qquad \text{(III.10)}$$

Where S denotes the source position $0 < S < l$ and $\lambda \in L^2(0,T)$ designates its time-dependent intensity function. Moreover, employing a source F of the form (III.10) implies that the problem (III.4)-(III.7) admits a unique solution u that belongs to :

$$L^2(0,T;H^1(0,l)) \cap C([0,T];L^2(0,l))$$

Chapitre III. Inverse source problem in a one-dimensional evolution linear transport equation with spatially varying coefficients : application to surface water pollution

Furthermore, assuming the time-dependent intensity function λ vanishes before reaching the final control time T which means

$$\exists T^* \in (0,T) \quad \text{such that} \quad \lambda(t) = 0 \quad \text{for all} \quad t \in (T^*,T) \tag{III.11}$$

Implies that $u(.,T^0) \in H^1(0,l)$ for all $T^0 \in (T^*,T)$. Then, we introduce the following Sturm-Liouville problem :

$$\begin{aligned}-\big(p(x)\psi'(x)\big)' + q(x)\psi(x) &= \mu\ell(x)\psi(x) \quad for \quad 0 < x < l \\ \psi(0) = \psi'(l) &= 0\end{aligned} \tag{III.12}$$

Where p and ℓ are the two functions given in (III.3). Since p, p', q and ℓ are continuous on $[0,l]$ while $p(x) > 0$ and $\ell(x) > 0$ on $[0,l]$, the system (III.12) is a regular Sturm-Liouville problem [14]. Therefore, the eigenvalues μ_n for $n \geq 1$ are real, simple and can be ordered such that $\mu_1 < \mu_2 < ...$ with $\lim\limits_{n \to \infty} \mu_n = \infty$. In addition, the normalized eigenfunctions ψ_n associated to the eigenvalues μ_n for $n \geq 1$ form a complete orthonormal family of $L^2_\ell(0,l)$

$$< \psi_n, \psi_m >_{L^2_\ell(0,l)} = \int_0^l \psi_n(x)\psi_m(x)\ell(x)dx = \begin{cases} 1 & \text{if } n = m \\ 0 & \text{otherwise} \end{cases} \tag{III.13}$$

And for each $f \in L^2_\ell(0,l)$, the series $\sum_{n\geq 1} < f, \psi_n >_{L^2_\ell(0,l)} \psi_n$ converges to f in $L^2_\ell(0,l)$.

Remark 2.1 *In [41], the author proved that if the function f belongs to $H^1(0,l)$ and satisfies the same boundary conditions i.e., $f(0) = f'(l) = 0$, then the expansion $\sum_{n\geq 1} < f, \psi_n >_{L^2_\ell(0,l)} \psi_n$ converges absolutely and uniformly to f in $[0,l]$.*

Besides, we remind the concept of a strategic point as introduced by El Jai and Pritchard in [20] and employed by the authors in [17, 19].

Definition 2.2 *A point x_0 of $(0,l)$ is called strategic with respect to a complete orthogonal family of continuous functions $\{\psi_n\}$ if $\psi_n(x_0) \neq 0$ for all n.*

Let a and b be two real numbers such that $0 < a < b < l$. For reasons to be explained later, we introduce the following two functions φ_1 and φ_2 which are the impulse response to the operator that is the adjoint of the spatial part of the operator L introduced in (III.5) :

III.2 Mathematical modelling and problem statement

$$\begin{aligned}-\bigl(p(x)\varphi_1'(x)\bigr)' + q(x)\varphi_1(x) = \delta(x-a) &\quad \text{and} \quad \varphi_1(0) = \varphi_1(b) = 0 \\ -\bigl(p(x)\varphi_2'(x)\bigr)' + q(x)\varphi_2(x) = \delta(x-b) &\quad \text{and} \quad \varphi_2(a) = \varphi_2'(l) = 0\end{aligned} \quad \text{(III.14)}$$

Then, we prove that under a reasonable condition on the spatially varying coefficients v, D and r, the function φ_1 introduced in (III.14) does not admit any root in the interval (a,b) :

Lemma 2.3 *Provided the coefficients D, v and r satisfy the following condition :*

$$4r + \frac{v^2(x)}{D(x)} + 2v'(x) + 2D''(x) \geq \frac{(D'(x))^2}{D(x)} \quad \text{for} \quad 0 < x < l \quad \text{(III.15)}$$

The function φ_1 introduced in (III.14) is such that $\varphi_1(x) \neq 0$ for all $x \in (a,b)$.

Proof. As the function p introduced in (III.3) is strictly positive on $[0,l]$, φ_1 satisfies :

$$\varphi_1''(x) + \frac{p'(x)}{p(x)}\varphi_1'(x) - \frac{q(x)}{p(x)}\varphi_1(x) = 0 \quad \text{for} \quad a < x < l \quad \text{(III.16)}$$

Then, using $\Phi_1(x) = \sqrt{p(x)}\varphi_1(x)$, the equation (III.16) is equivalent to

$$\Phi_1''(x) + g(x)\Phi_1(x) = 0 \quad \text{for} \quad a < x < l \quad \text{(III.17)}$$

Where the function g is defined, in view of (III.3), as follows :

$$\begin{aligned}g(x) &= -\frac{q(x)}{p(x)} - \frac{1}{2}\frac{p''(x)}{p(x)} + \frac{1}{4}\left(\frac{p'(x)}{p(x)}\right)^2 \\ &= -\frac{R(x)}{D(x)} - \frac{1}{4}\left(\frac{V(x)}{D(x)}\right)^2 + \frac{1}{2}\left(\frac{V(x)}{D(x)}\right)' \\ &= -\frac{1}{4D^2(x)}\left(4rD(x) + v^2(x) + 2v'(x)D(x) + 2D(x)D''(x) - \bigl(D'(x)\bigr)^2\right)\end{aligned} \quad \text{(III.18)}$$

Since $D(x) > 0$ for all x in $[0,l]$, then in view of the last equality in (III.18), the assertion (III.15) yields $g(x) \leq 0$ for all x in $(0,l)$. Therefore, as proved in [11], all solutions to (III.17) are nonoscillating solutions in (a,l). That implies Φ_1 has at most one root in (a,l). Furthermore,

Chapitre III. Inverse source problem in a one-dimensional evolution linear transport equation with spatially varying coefficients : application to surface water pollution

as φ_1 and Φ_1 have the same roots and $\varphi_1(b) = 0$, we conclude that $\varphi_1(x) \neq 0$ for all x in (a, b).
□

That leads to establish the following theorem :

Theorem 2.4 *If Lemma 2.3 applies, then the function Φ defined as follows :*

$$\Phi(x) = \frac{\varphi_2(x)}{\varphi_1(x)} \quad \text{for } x \in (a,b) \tag{III.19}$$

Is continuous and strictly monotonic.

Proof. In view of (III.14) and using Lemma 2.3, the function Φ introduced in (III.19) is smooth enough on (a, b) and we have

$$\Phi''(x) = \frac{\varphi_2''(x)\varphi_1(x) - \varphi_1''(x)\varphi_2(x)}{\varphi_1^2(x)} - 2\frac{\varphi_1'(x)}{\varphi_1(x)}\Phi'(x) \tag{III.20}$$

Besides, according to (III.14) we find

$$p(x)\Big(\varphi_2''(x)\varphi_1(x) - \varphi_1''(x)\varphi_2(x)\Big) = -p'(x)\Big(\varphi_2'(x)\varphi_1(x) - \varphi_1'(x)\varphi_2(x)\Big) \tag{III.21}$$

Which implies that

$$p(x)\frac{\varphi_2''(x)\varphi_1(x) - \varphi_1''(x)\varphi_2(x)}{\varphi_1^2(x)} = -p'(x)\Phi'(x) \tag{III.22}$$

Then, in view of (III.20) and (III.22), the function Φ satisfies in (a, b) the following second order differential equation :

$$p(x)\Phi''(x) + \Big(p'(x) + 2p(x)\frac{\varphi_1'(x)}{\varphi_1(x)}\Big)\Phi'(x) = 0 \tag{III.23}$$

Which leads to $\Phi'(x) = c/\big(p(x)\varphi_1(x)^2\big)$ where c is a real constant. Therefore, as according to (III.3) we have $p(x) > 0$ for all x in (a, b), it follows that the function Φ' has a fixed sign on (a, b). That implies Φ is a strictly monotonic function on (a, b). □

III.2 Mathematical modelling and problem statement

To establish the identifiability theorem, we need also to prove the following lemma :

Lemma 2.5 *Let $x_0 \in (0, l)$ be a strategic point with respect to the family $\{\psi_n\}$ as introduced in definition 2.2. If the solution w to the following system :*

$$\begin{aligned} L[w](x,t) &= 0 & 0 < x < l, \quad T^0 < t < T \\ w(.,T^0) &\in H^1(0,l) \\ w(0,t) &= \partial_x w(l,t) = 0 & T^0 < t < T \end{aligned} \qquad (III.24)$$

Satisfies $w(x_0, t) = 0$ for all t in (T^0, T), then we have $w(., T^0) = 0$ in $L_\ell^2(0, l)$.

Proof. Using the complete orthonormal family $\{\psi_n\}$, we express the solution w to the system (III.24) at the strategic point x_0 as follows :

$$w(x_0, t) = \sum_{n \geq 1} < w(., T^0), \psi_n >_{L_\ell^2(0,l)} e^{-\mu_n(t-T^0)} \psi_n(x_0) \qquad (III.25)$$

Then, since $w(., T^0)$ belongs to $H^1(0, l)$ and in view of Remark 4.2, it follows from the uniform convergence in $[0, l]$ of the expansion of $w(., T^0)$ in the complete family $\{\psi_n\}$ that in particular we have

$$\lim_{N \to \infty} \sum_{n=N}^{\infty} < w(., T^0), \psi_n >_{L_\ell^2(0,l)} \psi_n(x_0) = 0 \qquad (III.26)$$

Furthermore, (III.26) implies that the series occurring in the right-hand side of (III.25) converges uniformly in $[T^0, +\infty[$ and represents a real analytic function with respect to the variable t. That gives a sens to $w(x_0, t)$ for $t \geq T^0$. Therefore, as we have

$$w(x_0, t) = 0 \quad \text{for all } t \in (T^0, T) \qquad (III.27)$$

It follows by analytic continuation that

$$\sum_{n \geq 1} < w(., T^0), \psi_n >_{L_\ell^2(0,l)} e^{-\mu_n(t-T^0)} \psi_n(x_0) = 0 \quad \text{for all } t > T^0 \qquad (III.28)$$

Chapitre III. Inverse source problem in a one-dimensional evolution linear transport equation with spatially varying coefficients : application to surface water pollution

Then, by rewriting (III.28) as follows :

$$e^{-\mu_1(t-T^0)}\left(<w(.,T^0),\psi_1>_{L^2_\ell(0,l)}\psi_1(x_0) + \sum_{n\geq 2}<w(.,T^0),\psi_n>_{L^2_\ell(0,l)} e^{-(\mu_n-\mu_1)(t-T^0)}\psi_n(x_0)\right) = 0$$

And setting the limit when t tends to $+\infty$, we find $<w(.,T^0),\psi_1>_{L^2_\ell(0,l)}\psi_1(x_0) = 0$. Hence, by repeating the same principle for all $n \geq 2$, we obtain

$$<w(.,T^0),\psi_n>_{L^2_\ell(0,l)}\psi_n(x_0) = 0 \quad \text{for all } n \geq 1 \quad (\text{III.29})$$

Since x_0 is a strategic point with respect to the family $\{\psi_n\}$, we conclude in view of (III.29) that $<w(.,T^0),\psi_n>_{L^2_\ell(0,l)}= 0$ for all $n \geq 1$ which implies $w(.,T^0) = 0$ in $L^2_\ell(0,l)$. □

3 Identifiability

Provided the main condition (III.15) holds true, we prove in this section that assuming the time-dependent intensity function λ satisfies (III.11), the elements defining the source F introduced in (III.10) are uniquely determined from recording the state u solution to (III.4)-(III.7) at two observation points a and b framing the source region. Note that in the case of constant coefficients D, v and r the main condition (III.15) is equivalent to $4r + v^2/D \geq 0$ which is always fulfilled. Therefore, the following theorem can be seen as a generalization of the identifiability result obtained in [30] for the case of equations with constant coefficients :

Theorem 3.1 *Let $F_i(x,t) = \lambda_i(t)\delta(x - S_i)$ where λ_i is a positive function of $L^2(0,T)$ that satisfies (III.11) and S_i is such that $0 < a < S_i < b < l$, for $i = 1,2$. Provided the main condition (III.15) holds true and at least one of the two observation points a, b is strategic with respect to the complete orthonormal family $\{\psi_n\}$, we have*

$$M[F_1] = M[F_2] \quad \Longrightarrow \quad S_1 = S_2 \quad \text{and} \quad \lambda_1(t) = \lambda_2(t) \quad a.e. \text{ in } (0,T) \quad (\text{III.30})$$

Proof. Let u_i be the solution to the system (III.4)-(III.7) with the time-dependent point source $F_i(x,t) = \lambda_i(t)\delta(x - S_i)$, for $i = 1,2$. Then, the variable $w = u_2 - u_1$ satisfies

III.3 Identifiability

$$\begin{aligned} L[w](x,t) &= \lambda_2(t)\ell(S_2)\delta(x-S_2) - \lambda_1(t)\ell(S_1)\delta(x-S_1) &&\text{for } 0<x<l,\ 0<t<T \\ w(x,0) &= 0 &&\text{for } 0<x<l \\ w(0,t) &= 0 \quad \text{and} \quad \partial_x w(l,t) = 0 &&\text{for } 0<t<T \end{aligned} \quad (\text{III.31})$$

Since λ_1 and λ_2 satisfy (III.11), we obtain from multiplying the first equation in (III.31) by the function φ_1 solution to the first system in (III.14) and integrating by parts over $(0,T^0)\times(0,b)$ using Green's formula where $T^0 \in (T^*, T)$, then by the function φ_2 solution to the second system in (III.14) and integrating by parts over $(0,T^0) \times (a,l)$ using Green's formula that

$$\begin{aligned} \bar{\lambda}_2\ell(S_2)\varphi_1(S_2) - \bar{\lambda}_1\ell(S_1)\varphi_1(S_1) &= P_1 \\ \bar{\lambda}_2\ell(S_2)\varphi_2(S_2) - \bar{\lambda}_1\ell(S_1)\varphi_2(S_1) &= P_2 \end{aligned} \quad (\text{III.32})$$

Where $\bar{\lambda}_i = \int_0^{T^*} \lambda_i(t)dt$ for $i=1,2$ and the coefficients P_1, P_2 are such that

$$\begin{aligned} P_1 &= \int_0^b w(x,T^0)\ell(x)\varphi_1(x)dx + \int_0^{T^0} \left(w(a,t) + [w(x,t)p(x)\varphi_1'(x) - \partial_x w(x,t)p(x)\varphi_1(x)]_0^b \right) dt \\ P_2 &= \int_a^l w(x,T^0)\ell(x)\varphi_2(x)dx + \int_0^{T^0} \left(w(b,t) + [w(x,t)p(x)\varphi_2'(x) - \partial_x w(x,t)p(x)\varphi_2(x)]_a^l \right) dt \end{aligned} \quad (\text{III.33})$$

Furthermore, as in view of (III.14) we have $\varphi_1(0) = \varphi_1(b) = 0$, $\varphi_2(a) = \varphi_2'(l) = 0$ and according to (III.9), $M[F_1] = M[F_2]$ implies that

$$w(a,t) = w(b,t) = 0 \quad \text{for } 0<t<T \quad (\text{III.34})$$

Then the coefficients P_1 and P_2 introduced in (III.33) are reduced to

$$\begin{aligned} P_1 &= \int_0^b w(x,T^0)\ell(x)\varphi_1(x)dx \\ P_2 &= \int_a^l w(x,T^0)\ell(x)\varphi_2(x)dx \end{aligned} \quad (\text{III.35})$$

Besides, as (III.11) holds, then for $T^0 \in (T^*, T)$ the variable w satisfies in $(0,l) \times (T^0, T)$ a system similar to the problem (III.31) where the right-hand side of the first equation vanishes and the initial condition is $w(.,T^0) \in H^1(0,l)$. Then, assuming the observation point b to be

Chapitre III. Inverse source problem in a one-dimensional evolution linear transport equation with spatially varying coefficients : application to surface water pollution

strategic and using, in view of (III.34), $w(b,t) = 0$ in (T^0, T) we obtain by applying Lemma 4.1 with $x_0 = b$ that $w(., T^0) = 0$ in $L_\ell^2(0, l)$. Therefore, according to (III.35) that leads to find $P_1 = P_2 = 0$. In addition, since the main condition (III.15) holds, we have according to Lemma 2.3 that $\varphi_1(S_1) \neq 0$ and $\varphi_1(S_2) \neq 0$. Thus, using (III.32), we find

$$\frac{\varphi_2(S_2)}{\varphi_1(S_2)} = \frac{\varphi_2(S_1)}{\varphi_1(S_1)} \quad \Leftrightarrow \quad \Phi(S_2) = \Phi(S_1) \tag{III.36}$$

Where Φ is the function introduced in (III.19). From (III.36) and using Theorem 2.4, we obtain $S_2 = S_1$. Now, by setting $S_2 = S_1 = S$ we have

$$\begin{aligned} L[w](x,t) &= \big(\lambda_2(t) - \lambda_1(t)\big)\ell(S)\delta(x - S) && \text{for } 0 < x < l,\ 0 < t < T^* \\ w(x,0) &= 0 && \text{for } 0 < x < l \\ w(0,t) &= 0 \quad \text{and} \quad \partial_x w(l,t) = 0 && \text{for } 0 < t < T^* \end{aligned} \tag{III.37}$$

Then, using the complete orthonormal family $\{\psi_n\}$ to compute the solution w of (III.37) and the Titchmarsh's theorem on convolution of L^1 functions [60], we prove by employing similar techniques to those used in [17] that the assumptions b is a strategic point and $w(b,t) = 0$ for $0 < t < T^*$ imply that $\lambda_1(t) = \lambda_2(t)$ almost everywhere in $(0, T^*)$. □

4 Identification

In this section, we focus on establishing an identification method that uses the records (III.9) to determine the elements defining the source F introduced in (III.10). To this end, we proceed in two steps : a first step enables to localize the source position S and compute the mean value of the loaded intensity function λ. Then, a second step uses the determined source position and transforms the recovery of λ into solving a deconvolution problem.

4.1 Step1 : Localization of the source position S

Proposition 4.1 *Let F be a time-dependent point source as introduced in (III.10) where λ satisfies (III.11) and let $T^0 \in (T^*, T)$. Provided the coefficients D, v and r satisfy the main condition (III.15), the source position S and $\bar{\lambda} = \int_0^{T^*} \lambda(t)dt$ are subject to :*

III.4 Identification

$$\Phi(S) = \frac{Q_2}{Q_1} \quad \text{and} \quad \bar{\lambda} = \frac{Q_1}{\ell(S)\varphi_1(S)} \tag{III.38}$$

Where φ_1, Φ are the two functions introduced in (III.14), (III.19) and Q_1, Q_2 are such that

$$\begin{aligned} Q_1 &= \int_0^b u(x,T^0)\ell(x)\varphi_1(x)dx + \int_0^{T^0} \Big(u(a,t) + p(b)\varphi_1'(b)u(b,t)\Big)dt \\ Q_2 &= \int_a^l u(x,T^0)\ell(x)\varphi_2(x)dx + \int_0^{T^0} \Big(u(b,t) - p(a)\varphi_2'(a)u(a,t)\Big)dt \end{aligned} \tag{III.39}$$

Proof. Let u be the solution to (III.4)-(III.7) with the time-dependent point source F introduced in (III.10). Since (III.11) holds, we obtain from multiplying the equation (III.4) by the function φ_1 solution to the first system introduced in (III.14) and integrating by parts over $(0,T^0) \times (0,b)$ using Green's formula where $T^0 \in (T^*, T)$, then by the function φ_2 solution to the second system in (III.14) and integrating by parts over $(0,T^0) \times (a,l)$ using Green's formula that

$$\begin{aligned} \bar{\lambda}\ell(S)\varphi_1(S) &= Q_1 \\ \bar{\lambda}\ell(S)\varphi_2(S) &= Q_2 \end{aligned} \tag{III.40}$$

Where $\bar{\lambda} = \int_0^{T^*} \lambda(t)dt$ and the coefficients Q_1, Q_2 are given by

$$\begin{aligned} Q_1 &= \int_0^b u(x,T^0)\ell(x)\varphi_1(x)dx + \int_0^{T^0} \left(u(a,t) + \Big[u(x,t)p(x)\varphi_1'(x) - \partial_x u(x,t)p(x)\varphi_1(x)\Big]_0^b\right)dt \\ Q_2 &= \int_a^l u(x,T^0)\ell(x)\varphi_2(x)dx + \int_0^{T^0} \left(u(b,t) + \Big[u(x,t)p(x)\varphi_2'(x) - \partial_x u(x,t)p(x)\varphi_2(x)\Big]_a^l\right)dt \end{aligned} \tag{III.41}$$

Therefore, using the boundary conditions on u and $\varphi_1(0) = \varphi_1(b) = 0$, $\varphi_2(a) = \varphi_2'(l) = 0$ in (III.41), we find the coefficients Q_1 and Q_2 introduced in (III.39). Furthermore, since the main condition (III.15) holds we have according to Lemma 2.3 that $\varphi_1(S) \neq 0$. And thus, from (III.40) we obtain the result announced in (III.38). □

Remark 4.2 *Note that as u is subject to only knowledge of $u(a,t)$ and $u(b,t)$ for $0 < t < T$, the computation of the source position S and $\bar{\lambda}$ from (III.38) is not so far possible since the coefficients Q_1 and Q_2 derived in (III.39) still involve the unknown data $u(.,T^0)$.*

Chapitre III. Inverse source problem in a one-dimensional evolution linear transport equation with spatially varying coefficients : application to surface water pollution

To determine the two integrals in (III.39) involving the unknown data $u(.,T^0)$, we prove the following proposition :

Proposition 4.3 *Assuming (III.11) holds, let $T^0 \in (T^*, T)$ and μ_{n_0} be the eventual null eigenvalue of the regular Sturm-Liouville problem introduced on (III.12). Then, we have*

$$\begin{aligned} \int_0^b u(x,T^0)\ell(x)\varphi_1(x)dx &= \xi_{n_0} \int_0^b \ell(x)\varphi_1(x)\psi_{n_0}(x)dx + \sum_{1 \leq n \neq n_0} (\psi_n(a) + p(b)\varphi_1'(b)\psi_n(b)) \frac{\xi_n}{\mu_n} \\ \int_a^l u(x,T^0)\ell(x)\varphi_2(x)dx &= \xi_{n_0} \int_a^l \ell(x)\varphi_2(x)\psi_{n_0}(x)dx + \sum_{1 \leq n \neq n_0} (\psi_n(b) - p(a)\varphi_2'(a)\psi_n(a)) \frac{\xi_n}{\mu_n} \end{aligned} \quad \text{(III.42)}$$

Where $\xi_n = <u(.,T^0), \psi_n>_{L^2_\ell(0,l)}$ for all $n \geq 1$.

Proof. Since (III.11) holds, then for $T^0 \in (T^*, T)$ the solution u to the problem (III.4)-(III.7) is such that $u(.,T^0) \in H^1(0,l)$. Therefore, according to Remark 2.1, the series $\sum_{n \geq 1} \xi_n \psi_n$ with $\xi_n = <u(.,T^0), \psi_n>_{L^2_\ell(0,l)}$ for all $n \geq 1$ converges uniformly to $u(.,T^0)$ in $[0,l]$. And thus, using Lebesgue's theorem of dominated convergence, we obtain

$$\begin{aligned} \int_0^b u(x,T^0)\ell(x)\varphi_1(x)dx &= \xi_{n_0} \int_0^b \ell(x)\varphi_1(x)\psi_{n_0}(x)dx + \sum_{1 \leq n \neq n_0} \xi_n \int_0^b \ell(x)\psi_n(x)\varphi_1(x)dx \\ \int_a^l u(x,T^0)\ell(x)\varphi_2(x)dx &= \xi_{n_0} \int_a^l \ell(x)\varphi_2(x)\psi_{n_0}(x)dx + \sum_{1 \leq n \neq n_0} \xi_n \int_a^l \ell(x)\psi_n(x)\varphi_2(x)dx \end{aligned} \quad \text{(III.43)}$$

Then, multiplying the first equation in the regular Sturm-Liouville problem introduced in (III.12) firstly by the function φ_1 solution to the first system in (III.14) and integrating by parts using Green's formula over $(0,b)$, then by the function φ_2 solution to the second system in (III.14) and integrating by parts using Green's formula over (a,l), we find

$$\begin{aligned} \mu_n \int_0^b \ell(x)\psi_n(x)\varphi_1(x)dx &= \psi_n(a) + p(b)\varphi_1'(b)\psi_n(b) \\ \mu_n \int_a^l \ell(x)\psi_n(x)\varphi_2(x)dx &= \psi_n(b) - p(a)\varphi_2'(a)\psi_n(a) \end{aligned} \quad \text{(III.44)}$$

Hence, using (III.44) in (III.43) gives the result announced in (III.42). □

Note that as in view of (III.53) the eigenvalues μ_n for $n \geq 1$ are asymptotically quadratic with respect to n and all the coefficients ξ_n are bounded by $\|u(.,T^0)\|_{L^2_\ell(0,l)}$, Proposition 4.3

III.4 Identification

suggests that the series in (III.42) may be truncated based on a finite sufficiently large number N of initial terms. Furthermore, to determine the N coefficients ξ_n for $n = 1, .., N$ defining the truncated series in (III.42), we use the following system satisfied by u :

$$
\begin{aligned}
&L[u](x,t) = 0 &&\text{for} \quad 0 < x < l, \ T^0 < t < T \\
&u(.,T^0) \in H^1(0,l) \\
&u(0,t) = 0 \quad \text{and} \quad \partial_x u(l,t) = 0 &&\text{for} \quad T^0 < t < T
\end{aligned}
\tag{III.45}
$$

In addition, using the complete orthonormal family $\{\psi_n\}$, we approximate the solution u of the system (III.45) taken at the downstream observation point b as follows :

$$
u(b,t) \approx \sum_{n=1}^{N} \xi_n e^{-\mu_n(t-T^0)} \psi_n(b) \qquad \text{where} \quad \xi_n = <u(.,T^0), \psi_n>_{L^2_\ell(0,l)} \tag{III.46}
$$

Then, using the records d_b of the solution u taken at some discrete times t_m of the interval $[T^0, T]$ for $m = m_0, .., M_0$ where $M_0 > m_0 > 0$, we determine the N coefficients ξ_n from solving the following quadratic minimization problem :

$$
\min_{\xi \in \mathbb{R}^N} \frac{1}{2} \left\| E\xi - U_b \right\|_2^2 + \frac{\varepsilon^2}{2} \|\xi\|_2^2 \tag{III.47}
$$

Here, E is the rectangular matrix of entries $E_{mn} = e^{-\mu_n(t_m - T^0)} \psi_n(b)$ for $m_0 \leq m \leq M_0$, $1 \leq n \leq N$ and $U_b = \left(d_b^{m_0}, .., d_b^{M_0} \right)^\top$ where $d_b^m = d_b(t_m)$ for $m = m_0, .., M_0$. Moreover, as the measures are usually uncertain, we used in (III.47) a Tikhonov regularization term. The regularization parameter ε should be choosen as a good compromise between fulfilling the physical model and ensuring the stability of the computed solution. Thus, ε can be determined using Morozov's discrepancy principle, see for example [59, 62].

In order to solve the minimization problem (III.47), we need to determine the eigenpairs (μ_n, ψ_n) for $n = 1, .., N$. To this end, as in view of (III.3) we have $\ell(x)/p(x) = 1/D(x)$ and $p(x)\ell(x) = p^2(x)/D(x)$, we use the following change of variables : given x in $(0, l)$, let

$$
\eta = \int_0^x \frac{ds}{\sqrt{D(s)}} \qquad \text{and} \qquad \zeta(\eta) = \frac{\sqrt{p(x)}}{\left(D(x)\right)^{1/4}} \psi(x) \tag{III.48}
$$

That transforms the regular Sturm-Liouville problem introduced in (III.12) into the following equivalent Liouville normal form :

$$-\zeta''(\eta) + h(\eta)\zeta(\eta) = \mu\zeta(\eta) \quad \text{for} \quad 0 < \eta < I$$
$$\zeta(0) = 0 \quad \text{and} \quad \zeta'(I) + c\zeta(I) = 0 \tag{III.49}$$

Where $I = \int_0^l \left(D(s)\right)^{-1/2} ds$, the constant $c = \left(2v(l) - D'(l)\right)/\left(4\sqrt{D(l)}\right)$ and

$$h(\eta) = r + \frac{1}{2}v'(x) + \frac{1}{4}D''(x) + \frac{v^2(x)}{4D(x)} - \frac{\left(D'(x)\right)^2}{16D(x)} \tag{III.50}$$

Note that in view of the regularity of the coefficients D and v mentioned earlier in this paper, the function h introduced in (III.50) belongs to $L^2(0, I)$.

4.2 Step2 : Recovery of the time-dependent intensity function λ

In this section, we assume the source position S to be known and focus on recovering the history of the time-dependent intensity function λ. Then, assuming (III.11) holds and using the complete orthonormal family of eigenfunctions $\{\psi_n\}$, the solution u to the problem (III.4)-(III.7) with the time-dependent point source F introduced in (III.10) is given by

$$u(x,t) = \ell(S) \sum_{n \geq 1} \psi_n(S) \int_0^t \lambda(\eta) e^{-\mu_n(t-\eta)} d\eta \, \psi_n(x) \quad \text{for } (x,t) \in (0,l) \times (0,T^*) \tag{III.51}$$

Moreover, the solution u in (III.51) can be rewritten as follows :

$$u(x,t) = \int_0^t \lambda(\eta) \mathcal{K}(x, t-\eta) \, d\eta \quad \text{where} \quad \mathcal{K}(x,t) = \ell(S) \sum_{n \geq 1} \psi_n(S) \psi_n(x) e^{-\mu_n t} \tag{III.52}$$

Here, (III.52) is obtained from (III.51) by inversion of summation and integration. This inversion is justified by the Lebegues's theorem of dominated convergence : In fact according to [14, 53], as the function h introduced in (III.50) belongs to $L^2(0, I)$, the eigenvalues of (III.49) are simple and satisfy the following asymptotic result :

III.4 Identification

$$\mu_n = (n + \frac{1}{2})^2 \frac{\pi^2}{I^2} + 2c + \int_0^I h(\eta)d\eta + \theta_n \qquad (\text{III.53})$$

Where $\sum_{n\geq 1} \theta_n^2 < \infty$. Therefore, there exists $\hat{n} > 0$ and a real constant C_0 such that we have $\mu_n \geq n^2\pi^2/I^2 + C_0$ for all $n \geq \hat{n}$. Furthermore, since the eigenfunctions ψ_n for $n \geq 1$ are bounded in $[0, l]$ and the time variable t belongs to $(0, T^*)$, then there exists a positive real constant C for which we have

$$\begin{aligned}
\sum_{n\geq \hat{n}} |\psi_n(S)\psi_n(x)e^{-\mu_n t}| &\leq C \sum_{n\geq \hat{n}} e^{-(\pi/I)^2 t n^2} \\
&\leq C \int_0^{+\infty} e^{-(\pi\sqrt{t}/I)^2 s^2} ds = \frac{CI}{2\sqrt{\pi}} \frac{1}{\sqrt{t}}
\end{aligned} \qquad (\text{III.54})$$

In the remainder of this section, we focus on using (III.52) to recover the time-dependent intensity function λ. As the transport is naturally oriented downstream, it seems to be more convenient to use the downstream concentration records d_b rather than the upstream records d_a in order to identify λ. Given $M_* > 0$, let t_m for $m = 1, .., M_*$ be discrete times regularly distributed with the uniform time-step $\Delta t = T^*/M_* : t_m = m\Delta t$ for $m = 1, .., M_*$. Furthermore, we employ the following partial sum :

$$\mathcal{K}_N(b, t) = \ell(S) \sum_{n=1}^{N} \psi_n(S)\psi_n(b)e^{-\mu_n t} \qquad (\text{III.55})$$

As an approximation to the kernel $\mathcal{K}$ introduced in (III.52) at the downstream observation point b. Therefore, according to (III.52) we are interested in finding λ such that

$$\int_0^{t_m} \lambda(s)\mathcal{K}_N(b, t_m - s)ds = d_b^m \qquad \text{for} \quad m = 1, .., M_* \qquad (\text{III.56})$$

Where $d_b^m = d_b(t_m)$ for $m = 1, .., M_*$. In addition, using the trapezoidal rule, we get

Chapitre III. Inverse source problem in a one-dimensional evolution linear transport equation with spatially varying coefficients : application to surface water pollution

$$\int_0^{t_m} \lambda(s)\mathcal{K}_N(b, t_m - s)ds = \sum_{k=1}^{m} \int_{t_{k-1}}^{t_k} \lambda(s)\mathcal{K}_N(b, t_m - s)ds$$

$$\approx \frac{\Delta t}{2} \sum_{k=1}^{m} \mathcal{K}_N(b, t_m - t_k)\lambda_k + \mathcal{K}_N(b, t_m - t_{k-1})\lambda_{k-1} \quad \text{(III.57)}$$

$$= \Delta t \sum_{k=1}^{m-1} \mathcal{K}_N(b, t_m - t_k)\lambda_k + \frac{\Delta t}{2}\mathcal{K}_N(b, 0)\lambda_m$$

Where $\lambda_k = \lambda(t_k)$ for $k = 1, .., m$ and $\lambda(t_0) = 0$. Hence, we obtain the following discretized version of the problem (III.56) : find the vector $\Lambda = (\lambda_1, .., \lambda_{M_*})^\top$ in $I\!\!R^{M_*}$ such that

$$F^N \Lambda = U_b \qquad \text{where} \quad U_b = (d_b^1, .., d_b^{M_*})^\top \in I\!\!R^{M_*} \quad \text{(III.58)}$$

And F^N is the real lower triangular $M_* \times M_*$ matrix defined by

$$\begin{aligned} F_{mm}^N &= \frac{\Delta t}{2}\mathcal{K}_N(b, 0) & \text{for} \quad m = 1, .., M_* \\ F_{mk}^N &= \Delta t \mathcal{K}_N(b, t_m - t_k) & \text{for} \quad k = 1, .., m-1 \end{aligned} \quad \text{(III.59)}$$

Therefore, provided $\mathcal{K}_N(b, 0) \neq 0$, we deduce from the linear system introduced in (III.57) the following recursive formula that enables to determine the sought vector Λ :

$$\lambda_m = \frac{2}{\Delta t \mathcal{K}_N(b, 0)} \left(d_b^m - \Delta t \sum_{k=1}^{m-1} \lambda_k \mathcal{K}_N(b, t_m - t_k) \right) \quad \text{for} \quad m = 1, .., M_* \quad \text{(III.60)}$$

In the following proposition, we prove that we have $\mathcal{K}_N(b, 0) \neq 0$ for almost all $N \geq 2$:

Proposition 4.4 *Let b be a strategic point with respect to the complete orthonormal family of eigenfunctions $\{\psi_n\}$ and $S \in (0, l)$. For all $N \geq 2$, if $\mathcal{K}_N$ introduced in (III.55) is such that $\mathcal{K}_N(b, 0) = 0$, then at least one of the two real numbers $\mathcal{K}_{N+1}(b, 0)$ and $\mathcal{K}_{N+2}(b, 0)$ is different to zero.*

Proof. According to (III.3) and definition 2.2, we have $\ell(S) \neq 0$ for all S in $(0, l)$ and $\psi_n(b) \neq 0$ for all $n \geq 1$. Therefore, in view of (III.55) to achieve the proof we need only to show that for all $N \geq 2$ the two consecutive eigenfunctions ψ_{N+1} and ψ_{N+2} do not have any common zero in $(0, l)$. To this end, given $N \geq 2$ let ζ_{N+1} and ζ_{N+2} be the two eigenfunctions associate to the eigenvalues μ_{N+1} and μ_{N+2} of the Liouville normal form introduced in (III.49). Then, we have

$$\left(\zeta'_{N+1}\zeta_{N+2} - \zeta_{N+1}\zeta'_{N+2}\right)' = (\mu_{N+2} - \mu_{N+1})\zeta_{N+1}\zeta_{N+2} \qquad \text{(III.61)}$$

By integrating the equation given in (III.61) between two consecutive zeros η_0 and η_1 of the eigenfunction ζ_{N+1}, we obtain

$$(\mu_{N+2} - \mu_{N+1})\int_{\eta_0}^{\eta_1}\zeta_{N+2}(\eta)\zeta_{N+1}(\eta)d\eta = \zeta'_{N+1}(\eta_1)\zeta_{N+2}(\eta_1) - \zeta'_{N+1}(\eta_0)\zeta_{N+2}(\eta_0) \qquad \text{(III.62)}$$

Furthermore, we may assume $\zeta_{N+1}(\eta) > 0$ for $\eta_0 < \eta < \eta_1$ which implies that $\zeta'_{N+1}(\eta_0) > 0$ and $\zeta'_{N+1}(\eta_1) < 0$. Therefore, in view of (III.62) the function ζ_{N+2} should have a zero in the open interval (η_0, η_1). Otherwise, we get a contradiction between the two signs of the left and the right sides in (III.62). Moreover, using the same analysis, we prove that ζ_{N+2} has also a zero situated strictly between $\eta = 0$ and the first zero of ζ_{N+1} and another zero strictly between the last zero of ζ_{N+1} and $\eta = I$.

Consequently, as ζ_{N+1} and ζ_{N+2} have exactly N and $N+1$ zeros in $(0, I)$, we conclude that these two consecutive eigenfunctions do not have any common zero in $(0, I)$. Since in view of (III.3) and (III.48) there is a correspondence one by one between the zeros of the two functions ψ_n and ζ_n, then ψ_{N+1} and ψ_{N+2} do not have any common zero in $(0, l)$. $\square$

5 Numerical experiments

In this section, we start by deriving the undimensioned version of the considered problem. Then, we introduce a particular choice for the spatially varying diffusion, velocity and reaction coefficients. Some numerical experiments using the introduced coefficients are carried out. We end this section by analysing the obtained numerical results and pointing out an outlook for the present study related to the Peclet number.

5.1 Undimensioned Problem

To derive the undimensioned version of the considered problem, we introduce the variables $(y, s) \in (0, 1)^2$ such that given $(x, t) \in (0, l) \times (0, T)$ associates $y = x/l$ and $s = t/T$. Then, we

Chapitre III. Inverse source problem in a one-dimensional evolution linear transport equation with spatially varying coefficients : application to surface water pollution

use the following notations :

$$\tilde{u}(y,s) = u(x,t) \quad \text{and} \quad \tilde{D}(y) = D(x), \quad \tilde{v}(y) = v(x) \tag{III.63}$$

Let $s^* = T^*/T$ and $\tilde{\lambda}(s) = \lambda(t)$. Thus, assuming (III.11) holds, we have $\lambda(s) = 0$ for all $s \geq s^*$. Therefore, the reduced state $\tilde{u}$ satisfies

$$\begin{aligned}
&\tilde{L}[\tilde{u}](y,s) = T\tilde{\lambda}(s)\tilde{\ell}(\tilde{S})\delta(y-\tilde{S}) && \text{for } 0 < y < 1, \;\; 0 < s < s^* \\
&\tilde{u}(y,0) = 0 && \text{for } 0 < y < 1 \\
&\tilde{u}(0,s) = 0 \;\; \text{and} \;\; \partial_y u(1,s) = 0 && \text{for } 0 < s < s^*
\end{aligned} \tag{III.64}$$

Where $\tilde{S} = S/l$ and $\tilde{L}[\tilde{u}](y,s) = \tilde{\ell}(y)\partial_s \tilde{u}(y,s) - \partial_y\big(\tilde{p}(y)\partial_y \tilde{u}(y,s)\big) + \tilde{q}(y)\tilde{u}(y,s)$. Here, $\tilde{q}(y) = \tilde{\ell}(y)\tilde{R}(y)$ with $\tilde{V}(y) = \frac{T}{l}\big(\tilde{v}(y) - \frac{\tilde{D}'(y)}{l}\big)$, $\tilde{R}(y) = T\big(r + \frac{\tilde{v}'(y)}{l}\big)$ and

$$\tilde{\ell}(y) = \frac{l^2}{T}\frac{\tilde{p}(y)}{\tilde{D}(y)} \quad \text{where} \quad \tilde{p}(y) = e^{-\frac{l^2}{T}\int_0^y \frac{\tilde{V}(\eta)}{\tilde{D}(\eta)}d\eta} \tag{III.65}$$

That reduces the Liouville normal form introduced in (III.49) to the following system :

$$\begin{aligned}
&-\tilde{\zeta}''(\tilde{\eta}) + \tilde{h}(\tilde{\eta})\tilde{\zeta}(\tilde{\eta}) = \mu\tilde{\zeta}(\tilde{\eta}) && \text{for } 0 < \tilde{\eta} < \tilde{I} \\
&\tilde{\zeta}(0) = 0 \;\; \text{and} \;\; \tilde{\zeta}'(\tilde{I}) + \tilde{c}\tilde{\zeta}(\tilde{I}) = 0
\end{aligned} \tag{III.66}$$

Where $\tilde{I} = \frac{1}{\sqrt{T}}\int_0^1 \tilde{D}^{-1/2}(z)dz$, the constant $\tilde{c} = \frac{\sqrt{T}}{4\sqrt{\tilde{D}(1)}}\big(2\tilde{v}(1) - \frac{\tilde{D}'(1)}{l}\big)$ and

$$\tilde{h}(\tilde{\eta}) = Tr + \frac{T}{2l}\tilde{v}'(y) + \frac{T}{4l^2}\tilde{D}''(y) + T\frac{\tilde{v}^2(y)}{4\tilde{D}(y)} - \frac{T}{16l^2}\frac{\big(\tilde{D}'(y)\big)^2}{\tilde{D}(y)} \tag{III.67}$$

With $\tilde{\eta} = \frac{1}{\sqrt{T}}\int_0^y \tilde{D}^{-1/2}(z)dz$. Then, given $N > 0$, we discretize the interval $[0, \tilde{I}]$ using the step size $\Delta\tilde{\eta} = \tilde{I}/(N+1)$ to obtain the regularly distributed $\tilde{\eta}_i = i\Delta\tilde{\eta}$ for $i = 0, .., N+1$. Furthermore, to compute the N eigenpairs $(\mu_n, \tilde{\zeta}_n)$ for $n = 1, .., N$ solutions to (III.66), we employ the three-point finite difference scheme with Numerov method [61]. That leads to the following generalized eigenproblem :

III.5 Numerical experiments

$$A\tilde{\zeta} = \mu B \tilde{\zeta} \quad \text{where} \quad A = \frac{1}{(\Delta\tilde{\eta})^2} M + B\tilde{H} \quad \text{and} \quad B = Id - \frac{1}{12} M \quad \text{(III.68)}$$

With Id is the identity matrix, $\tilde{H} = diag\big(\tilde{h}(\tilde{\eta}_1),..,\tilde{h}(\tilde{\eta}_N)\big)$ and

$$M = \begin{pmatrix} 2 & -1 & 0 & . & . & & 0 \\ -1 & 2 & -1 & 0 & . & & 0 \\ . & . & . & . & . & & . \\ 0 & . & 0 & -1 & 2 & & -1 \\ 0 & . & . & 0 & -1 & 2 - \dfrac{1}{1+\tilde{c}\Delta\tilde{\eta}} \end{pmatrix} \quad \text{(III.69)}$$

5.2 Particular choice for the coefficients D, v and r

To carry out numerical experiments, we use the following diffusion, velocity and reaction coefficients, see [51] :

$$D(x) = d_m + \alpha\big(1 - e^{-\beta x}\big), \quad v(x) = v_0 \quad \text{and} \quad r = r_0 \quad \text{(III.70)}$$

Where $d_m > 0$ is the molecular diffusion coefficient and α, β, v_0 and r_0 are four positive real numbers. Then, using (III.70) and the change of variable $\tilde{w}(y,s) = e^{Tr_0 s}\tilde{u}(y,s)$, the system (III.64) is rewritten as follows :

$$\begin{aligned}
&\tilde{\ell}(y)\partial_s \tilde{w}(y,s) - \partial_y\big(\tilde{p}(y)\partial_y \tilde{w}(y,s)\big) = Te^{Tr_0 s}\tilde{\lambda}(s)\tilde{\ell}(\tilde{S})\delta(y - \tilde{S}) && \text{for } (y,s) \in (0,1) \times (0,s^*) \\
&\tilde{w}(y,0) = 0 && \text{for } 0 < y < 1 \\
&\tilde{w}(0,s) = 0 \quad \text{and} \quad \partial_y \tilde{w}(1,s) = 0 && \text{for } 0 < s < s^*
\end{aligned} \quad \text{(III.71)}$$

Therefore, as introduced in (III.14), the functions $\tilde{\varphi}_1$ and $\tilde{\varphi}_2$ associated to (III.71) are such that

$$\begin{aligned}
-\big(\tilde{p}(y)\tilde{\varphi}_1'(y)\big)' &= \delta(y - \tilde{a}) && \text{and} \quad \tilde{\varphi}_1(0) = \tilde{\varphi}_1(\tilde{b}) = 0 \\
-\big(\tilde{p}(y)\tilde{\varphi}_2'(y)\big)' &= \delta(y - \tilde{b}) && \text{and} \quad \tilde{\varphi}_2(\tilde{a}) = \tilde{\varphi}_2'(1) = 0
\end{aligned} \quad \text{(III.72)}$$

Chapitre III. Inverse source problem in a one-dimensional evolution linear transport equation with spatially varying coefficients : application to surface water pollution

Where $\tilde{a} = a/l$ and $\tilde{b} = b/l$ are the two undimensioned observation points such that : $0 < \tilde{a} < \tilde{S} < \tilde{b} < 1$. Then, from solving the system (III.72) we obtain

$$\tilde{\varphi}_1(y) = \left(c_1 - \mathcal{H}(y - \tilde{a})\right) \int_{\tilde{a}}^{y} \tilde{p}^{-1}(\eta) d\eta + c_1 \int_{0}^{\tilde{a}} \tilde{p}^{-1}(\eta) d\eta$$
$$\tilde{\varphi}_2(y) = \left(1 - \mathcal{H}(y - \tilde{b})\right) \int_{\tilde{b}}^{y} \tilde{p}^{-1}(\eta) d\eta + \int_{\tilde{a}}^{\tilde{b}} \tilde{p}^{-1}(\eta) d\eta \quad \text{(III.73)}$$

With $c_1 = \int_{\tilde{a}}^{\tilde{b}} \tilde{p}^{-1}(\eta) d\eta / \int_{0}^{\tilde{b}} \tilde{p}^{-1}(\eta) d\eta$ and $\mathcal{H}$ is the Heaviside function [57]. Furthermore, according to (III.65) and using (III.70), we find

$$\tilde{p}(y) = e^{\left(-\frac{lv_0}{d_m + \alpha}y + \left(\frac{v_0}{\beta(d_m + \alpha)} - 1\right)\log\left(\frac{d_m}{d_m + \alpha(1 - e^{-\beta ly})}\right)\right)} \quad \text{for } 0 \le y \le 1 \quad \text{(III.74)}$$

Hence, for $s^0 = T^0/T$ and as established in Proposition 4.1, multiplying the first equation in (III.71) by $\tilde{\varphi}_1$ and integrating by parts over $(0, \tilde{b}) \times (0, s^0)$, then by $\tilde{\varphi}_2$ and integrating by parts over $(\tilde{a}, 1) \times (0, s^0)$ gives, since $\tilde{p}(\tilde{a})\tilde{\varphi}_2'(\tilde{a}) = 1$ and $\tilde{p}(\tilde{b})\tilde{\varphi}_1'(\tilde{b}) = (c_1 - 1)$, that

$$\tilde{\Phi}(\tilde{S}) = \tilde{Q} \quad \text{where} \quad \tilde{Q} = \frac{\int_{\tilde{a}}^{1} \tilde{w}(y, s^0)\tilde{\ell}(y)\tilde{\varphi}_2(y) dy + \int_{0}^{s^0} \left(\tilde{w}(\tilde{b}, s) - \tilde{w}(\tilde{a}, s)\right) ds}{\int_{0}^{\tilde{b}} \tilde{w}(y, s^0)\tilde{\ell}(y)\tilde{\varphi}_1(y) dy + \int_{0}^{s^0} \left(\tilde{w}(\tilde{a}, s) + (c_1 - 1)\tilde{w}(\tilde{b}, s)\right) ds} \quad \text{(III.75)}$$

And $\tilde{\Phi}(\tilde{S}) = \tilde{\varphi}_2(\tilde{S})/\tilde{\varphi}_1(\tilde{S})$. Furthermore, to determine the reduced source position $\tilde{S}$ from (III.75), we need to prove the following result :

Proposition 5.1 *Provided the diffusion parameters d_m, α, β and the velocity v_0 satisfy*

$$v_0 \ge \alpha\beta\sqrt{e^{-\beta ly}\left(2\left(1 + \frac{d_m}{\alpha}\right) - e^{-\beta ly}\right)} \quad \text{for all } 0 < y < 1 \quad \text{(III.76)}$$

The function $\tilde{\Phi} : y \mapsto \tilde{\varphi}_2(y)/\tilde{\varphi}_1(y)$ is well defined on $(\tilde{a}, \tilde{b})$ and we have

$$\tilde{\Phi}(y) = \tilde{\Phi}_0 \frac{\gamma(y) - \gamma(\tilde{a})}{\gamma(\tilde{b}) - \gamma(y)} \quad \text{where} \quad \tilde{\Phi}_0 = \frac{\gamma(\tilde{b}) - d_m^{\frac{v_0}{\beta(\alpha + d_m)}}}{\gamma(\tilde{a}) - d_m^{\frac{v_0}{\beta(\alpha + d_m)}}} \quad \text{(III.77)}$$

III.5 Numerical experiments

And $\gamma(z) = \left((d_m + \alpha)e^{\beta l z} - \alpha\right)^{\frac{v_0}{\beta(\alpha+d_m)}}$ for all $z \in [\tilde{a}, \tilde{b}]$.

Proof. See the appendix.

Therefore, from (III.75) and using (III.77), we determine the reduced source position $\tilde{S}$ as follows :

$$\tilde{S} = \frac{1}{\beta l} \ln\left(\frac{1}{d_m + \alpha}\left[\alpha + \left(\frac{\tilde{Q}\gamma(\tilde{b}) + \tilde{\Phi}_0\gamma(\tilde{a})}{\tilde{\Phi}_0 + \tilde{Q}}\right)^{\frac{\beta(d_m+\alpha)}{v_0}}\right]\right) \quad \text{(III.78)}$$

Besides, to compute the diagonal matrix $\tilde{H}$ occurring in (III.68) using the function $\tilde{h}$ introduced in (III.67), we need to prove the following result that expresses y as a function of the variable $\tilde{\eta}$:

Proposition 5.2 Let d_m, α and β be three real positive numbers and for all $0 \leq y \leq 1$, $\tilde{D}(y) = d_m + \alpha\left(1 - e^{-\beta l y}\right)$. Then, $\tilde{\eta} = \frac{l}{\sqrt{T}} \int_0^y \frac{dz}{\sqrt{\tilde{D}(z)}}$ is equivalent to

$$y = \frac{\sqrt{T(d_m + \alpha)}}{l}\tilde{\eta} + \frac{2}{\beta l}\ln\left(\frac{d_m\left(1 + \sqrt{1 + \frac{\alpha}{d_m}}\right) + \frac{\alpha}{2}\left(1 + e^{-\beta\sqrt{T(d_m+\alpha)}\tilde{\eta}}\right)}{(d_m + \alpha)\left(1 + \sqrt{\frac{d_m}{d_m+\alpha}}\right)}\right) \quad \text{(III.79)}$$

Proof. See the appendix.

To compute the N eigenpairs $(\mu_n, \tilde{\zeta}_n)$ for $n = 1, .., N$ solutions to the generalized eigenproblem introduced in (III.68)-(III.69), we used the function "bdiag" of the package Scilab to solve the ordinary eigenvalue problem : $B^{-1}A\tilde{\zeta} = \mu\tilde{\zeta}$. Then, according to (III.48) and (III.79), we deduce the eigenfunction $\tilde{\psi}_n$ associated to the computed eigenpair $(\mu_n, \tilde{\zeta}_n)$ as follows :

$$\tilde{\psi}_n(y_i) = \frac{\left(T\tilde{D}(y_i)\right)^{1/4}}{\sqrt{l\tilde{p}(y_i)}} \tilde{\zeta}_n(\tilde{\eta}_i) \qquad \text{for} \quad i = 1, .., N \quad \text{(III.80)}$$

Where for $i = 1, .., N$, y_i is the value associated to $\tilde{\eta}_i$ computed from (III.79). Therefore, by choosing the two observation points such that $\tilde{a} = y_{i_0}$ and $\tilde{b} = y_{i_1}$ where $1 \leq i_0 < i_1 \leq N$, we determine $\tilde{\psi}_n(\tilde{a})$ and $\tilde{\psi}_n(\tilde{b})$ for $n = 1, .., N$ using (III.80). Here, i_0 has to be taken small enough

Chapitre III. Inverse source problem in a one-dimensional evolution linear transport equation with spatially varying coefficients : application to surface water pollution

to keep $\tilde{a}$ in the upstream part of the reduced interval $(0, 1)$ and i_1 big enough to have $\tilde{b}$ in its downstream part as required by the identifiability theorem 3.1

5.3 Numerical tests and discussion

In this subsection, we use the established identification method to carry out some numerical experiments. To this end, we employ in (III.70) the following coefficients :

$$d_m = 10^{-5}\ m^2 s^{-1}, \quad \alpha = 10, \quad \beta = 10^{-2}, \quad v_0 = 0.1\ ms^{-1} \quad \text{and} \quad r_0 = 2 \times 10^{-3}\ s^{-1}$$

Then, we aim to identify the elements S and λ defining a sought time-dependent point source $F(x,t) = \lambda(t)\delta(x - S)$ occurring in the controlled portion of a river represented by the segment $(0, l)$ with $l = 1000m$. We assume controlling this portion of a river for $T = 14400s$ (4 hours) and $T^* = 10800s$ (3 hours). To generate the records d_a and d_b at the two observation points a and b, we solve the problem (III.4)-(III.7) with a source located at $S = 644m$ loading the following time-dependent intensity function :

$$\lambda(t) = \sum_{i=1}^{2} c_i e^{-v_i(t-\tau_i)^2} \quad \text{(III.81)}$$

Where $c_1 = 0.4$, $c_2 = 0.8$, $v_1 = 3 \times 10^{-7}$, $v_2 = 9 \times 10^{-7}$ and $\tau_1 = 4500$, $\tau_2 = 8000$.

Over the whole control time T, we employ $M_{Total} = 60$ measures of u at each of the two observation points a and b. Those measures have been taken at the regularly distributed discrete times $t_m = m\Delta t$ for $m = 1, .., M_{Total}$ where $\Delta t = T/M_{Total}$. Then, we have $T^* = M_*\Delta t$ with $M_* = 45$. We denote $d_a^m = d_a(t_m)$ and $d_b^m = d_b(t_m)$ the measures obtained from the records d_a and d_b taken at the discrete times t_m for $m = 1, .., M_{Total}$.

According to the identification method established in the previous section, we localize the source position S and recover λ by proceeding in the two following steps :

Step 1 : Set $T^0 = T^*$, $m_0 = M_*$, $M_0 = m_0 + N - 1$ and use the vector of measures $U_b = (d_b^{m_0}, .., d_b^{M_0})^\top$ to compute the N coefficients ξ_n from solving the quadratic minimization problem introduced in (III.47). To this end, we employ the conjugate gradient method. Then, we use the identified ξ_n for $n = 1, .., N$ to calculate the two unknown integrals $\int_0^b u(x, T^0)\ell(x)\varphi_1(x)dx$ and $\int_a^l u(x, T^0)\ell(x)\varphi_2(x)dx$ as established in Proposition 4.3 Furthermore, by employing the

III.5 Numerical experiments

trapezoidal rule and the measures d_a^m, d_b^m for $m = 1, .., M_*$, we compute $\int_0^{T^*} u(a,t) dt$ and $\int_0^{T^*} u(b,t) dt$. Therefore, we identify the source position S_{ident} and $\bar{\lambda}$ as given in Proposition 4.1

Step 2 : Use the identified source position $S = S_{ident}$ and check whether with the used N we have $\mathcal{K}_N(b, 0)$ introduced in (III.55) is not null. Otherwise, change the value of N according to Proposition 4.4 Then, set $S = S_{ident}$ and use the vector of measures $U_b = (d_b^1, ..., d_b^{M_*})^\top$ to calculate the unknown intensity vector $\Lambda_{ident} = \left(\lambda_1, .., \lambda_{M_*}\right)$ as derived in (III.60).

In the remainder, we are interested in studying numerically : how does the introduction of a noise on the used measures taken at the two observation points a and b affect the identified source elements. We carry out numerical experiments with $N = 15$, $i_0 = 4$ which corresponds to the upstream observation point $a = 157m$ and $i_1 = 14$ which corresponds to the downstream observation point $b = 857m$. Then, for each intensity of the introduced noise, we compute the relative error on the identified source intensity vector Λ_{ident} using

$$Error Lam = \frac{\|\Lambda - \Lambda_{ident}\|_2}{\|\Lambda\|_2} \quad \text{(III.82)}$$

Where $\Lambda = \left(\lambda(t_1), .., \lambda(t_{M_*})\right)$ with $\lambda(t)$ is the function introduced in (III.81) and $\|.\|_2$ represents the Euclidean norm. The results of this numerical study are presented below for different intensities of noise. For each case, we give the value of the identified source position S_{ident} and draw on the same figure the two curves showing the used intensity function introduced in (III.81) and the identified intensity function obtained from Λ_{ident}. We also give the relative error $ErrorLam$ computed using (III.82).

The analysis of the numerical experiments presented in figures $1-4$ shows that the established identification method enables to identify the elements defining the sought time-dependent point source with a relatively good accuracy. Those numerical results seem to be accurate and relatively stable with respect to the introduction of noises on the used measures.

Furthermore, according to **Step 2** we determine the source intensity function using the already identified source position S_{ident} in **Step 1**. Hence, a part of the error $ErrorLam$ on the identified source intensity function comes from the error already committed on the computed source position. In practice, usually the suspect sources of pollution are rather known and we are interested in identifying the responsable of the observed pollution. Therefore, **Step 1** could lead to deduce the exact position S of the sought source. Then, using S rather than S_{ident} in **Step 2** will improve the error $ErrorLam$.

Chapitre III. Inverse source problem in a one-dimensional evolution linear transport equation with spatially varying coefficients : application to surface water pollution

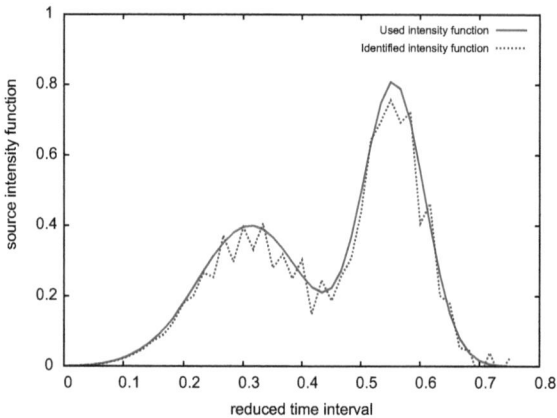

Figure 1. Noise intensity 3%: $S_{ident} = 635.91\ m$ and ErrorLam = 13.43%.

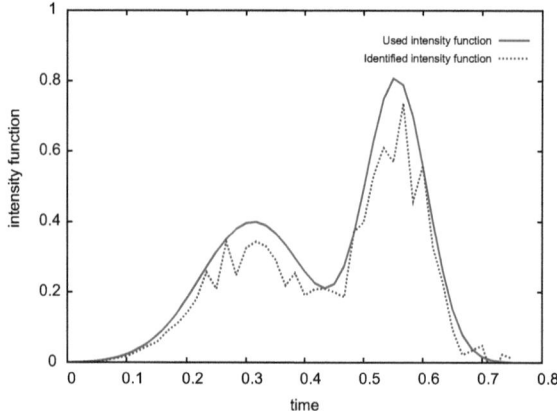

Figure 2. Noise intensity 5%: $S_{ident} = 620.45\ m$ and ErrorLam = 23.53%.

5.3.1. Discussion and outlook

Note that the results established in this paper require the use of two observation points (sensors) which should frame the source region. The importance of this requirement can be easily seen from the stationary version of the underlined inverse source problem: In that case, we aim to identify the two unknown parameters S and λ given some data measured by sensors. Then, for an explicit identification and since we have two unknowns, we need two measures. Furthermore, the analytic computation of the state reveals that it contains a term involving the Heaviside function at S. Therefore, if the two needed measures are taken at the same side of the source position (both upstream or both downstream) then, the

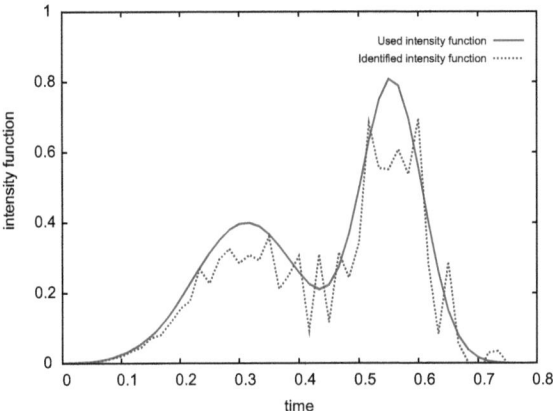

Figure 3. Noise intensity 7%: $S_{ident} = 666.2\ m$ and ErrorLam = 35.27%.

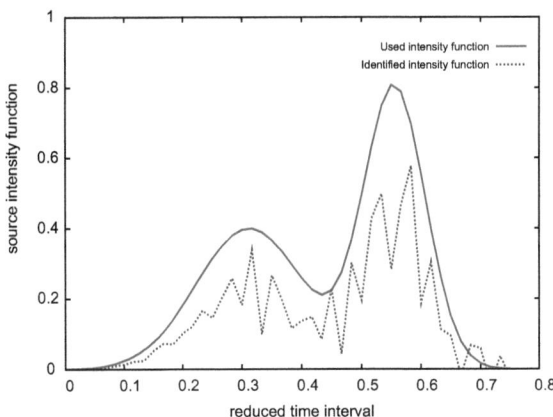

Figure 4. Noise intensity 10%: $S_{ident} = 590.65\ m$ and ErrorLam = 46.79%.

identification problem is equivalent to solving a system of two linearly dependent equations. Those two equations become linearly independent if one measure is taken upstream whereas the other is taken downstream with respect to the source position. That explains the need for two sensors and the fact that they should frame the source position. In practice, that seems to make sense since observing the source activity from only one side of the river will enable us to see some variation of the concentration but certainly not the significant change of its value between upstream and downstream regions. This significant change in the concentration represents the main characterization of the sought source.

As a work in progress, we are studying the robustness of the established identification method with respect to higher values of the Peclet number. This dimensionless number

Chapitre III. Inverse source problem in a one-dimensional evolution linear transport equation with spatially varying coefficients : application to surface water pollution

measures the ratio of the rate of advection by the rate of diffusion. The higher Peclet numbers correspond to the case of advection dominant flow. In such kind of flow, the damping effect exerted by the diffusion will be reduced and thus, from the engineering point of view one expects more sensitivity on the signals recorded by sensors. Another interesting point of this work in progress is how to select the total number of discrete times M_{Total}. The value $M_{Total} = 60$ used in this paper was selected after numerous runs as the value of M_{Total} from which the accuracy of the identified results does not improve significantly anymore. According to our first observations, the value of M_{Total} seems depending on the Peclet number and thus on the nature of the flow.

6 Conclusion

In this paper, we studied the identification of a time-dependent point source occurring in the right-hand side of a one-dimensional evolution linear transport equation with spatially varying diffusion, velocity and reaction coefficients. Under some reasonable conditions on those spatially varying coefficients and assuming the source intensity function vanishes before reaching the final control time, we proved the identifiability of the elements defining the sought time-dependent point source from recording the state at two observation points framing the source region. Then, we established an identification method that uses those records to localize the source position as the zero of a continuous and strictly monotonic function and transforms the task of recovering its intensity function into solving a deconvolution problem. Some numerical experiments on a variant of the water BOD pollution model are presented. The analysis of those experiments shows that the established identification method is accurate and stable with respect to the introduction of noises on the used measures.

Chapitre IV

Inverse source problem based on two dimensionless dispersion-current functions in 2D evolution transport equations with spatially varying coefficients

Les résultats de ce chapitre sont issus de l'article [36], soumis pour publication en collaboration avec Adel Hamdi.

1 Introduction

While dealing with inverse source problems, it is widely known that one of the main encountered difficulties is the no identifiability (uniqueness) in general of a source in its abstract form, see [34] for a counterexample. In the literature, to overcome this difficulty authors generally assume available some *a priori* information on the form of the sought source : For example, time-independent sources are treated by Cannon J.R. in [12] using spectral theory, then by Engl H., Scherzer O. and Yamamoto M. in [21] using the approximated controllability of the heat equation. The results of this last paper are generalized by M. Yamamoto in [66] to sources of separated time and space variables where the time-dependent part is assumed to be known and null at the initial time then, recently in [58] to sources where the known time-dependent part of the sought source could also depend on the space variables and the involved differential operator is a time fractional parabolic equation. Hettlich F. and Rundell W. addressed in [38] the $2D$ inverse source problem for the heat equation with as a source the characteristic function associated to a subset of a disk. They proved identifiability from recording the flux at two different points of the boundary. El Badia and Hamdi A. treated in [17, 30] the case

Chapitre IV. Inverse source problem based on two dimensionless dispersion-current functions in 2D evolution transport equations with spatially varying coefficients

of a time-dependent point source occurring in $1D$ evolution transport equations with constant coefficients. Recently, in [35] Hamdi A. and Mahfoudhi I. extended this study to the case of $1D$ evolution transport equations with spatially varying coefficients. Hamdi A. treated in [34] the case of $2D$ evolution advection-dispersion-reaction equations with *mean* velocity field and dispersion tensor. El Badia et al. studied in [3] the case of multiple time-dependent moving point sources in $2D$ evolution advection-diffusion-reaction equations with *constant* coefficients. The identifiability results established in [3, 34] are based on the unique continuation Theorem that requires only observations taken at any non-empty part of the boundary. That seems to be an ideal theoretical frame which in pratice doesn't take into account the flow properties, for example, advection dominant which implies that using only data reaching back the inflow boundary could not give a full information about the involved sources, or the so-called no-slipping condition which would suggest the same thing for using only data reaching the two lateral solid boundaries. Moreover, the identification method proposed in [3] consists in minimizing two classic objective functions (least squares and Kohn-Vogelius) defined from observations taken on the whole boundary.

In the present paper, we address the underlined nonlinear inverse source problem in $2D$ evolution advection-dispersion-reaction equations with spatially varying velocity field and dispersion tensor. **The originality** of this study consists in establishing a *constructive* identifiability theorem that leads to develop an identification method using *only* significant boundary observations and operating *other* than the classic minimization approach. That marks the difference between the results established in this paper and those in [3] for the case of constant velocity and diffusion coefficients. Besides, comparing the present study to [34] for the case of $2D$ transport equations with mean velocity field and dispersion tensor, we distinguish the three following points : **1.** The identifiability results proved in [3, 34] are based on the unique continuation Theorem which doesn't give any visibility on how to proceed in terms of identification whereas the identifiability theorem established in the present study is constructive and leads to develop an identification method using the same techniques **2.** The identification method proposed in [34] establishes only a linear link between the two coordinates of the sought source position. In the present study, we localize the source position as the unique solution of a nonlinear system of two equations and thus, we determine uniquely both coordinates of the sought position **3.** In this study, we derive lower and upper frame bounds for the total amount loaded by the source intensity function. As far as the identification of this latest is concerned, we prove a technical lemma based on some boundary controllability techniques that maintains the solution of an associated adjoint problem taken at an arbitrary spatial interior point as a non-null time-dependent function.

A motivation for our study is a typical problem associated with environmental monitoring that consists of identifying pollution sources in surface water. In a river, for example, the identification can be done by monitoring the BOD (Biochemical Oxygen Demand) concentration which represents the amount of dissolved oxygen consumed by the microorganisms during the oxidation process, see [13] for more details. The mathematical modelling of the BOD concentration is given by an evolution advection-dispersion-reaction equation [5, 43, 48, 50]. In the present study, we aim to identify a time-dependent point source occurring in a monitored portion of a river using some boundary records related to the BOD concentration. The paper is organized as follows : section 2 is devoted to stating the problem, assumptions and proving some technical results for later use. In section 3, we prove the identifiability of the elements defining the sought source from some boundary records related to the associated state. Section 4 is reserved to establish a constructive identification method that uses the required boundary records to determine the sought elements. We end this section by presenting an algorithm that summarizes the different steps of the established identification method. Some numerical experiments on a variant of the surface water BOD pollution model are given in section 5.

2 Problem statement

Let $T > 0$ be a final monitoring time and Ω be a bounded and simply connected open subset of $I\!R^2$ with a sufficiently smooth boundary $\partial\Omega := \Gamma_D \cup \Gamma_N$. Here, Γ_D denotes the inflow boundary of the domain Ω whereas Γ_N regroups its two lateral boundaries Γ_L and its outflow boundary Γ_{out} i.e., $\Gamma_N := \Gamma_L \cup \Gamma_{out}$. The BOD concentration, denoted here by u, is governed by the following equation [13, 48, 49] :

$$L[u](x,t) = F(x,t) \qquad \text{for all } (x,t) \in \Omega \times (0,T) \qquad (IV.1)$$

Where F represents the pollution source and L is the second-order linear parabolic partial differential operator defined as follows :

$$L[u](x,t) := \partial_t u(x,t) - \text{div}\left(D(x)\nabla u(x,t)\right) + V(x)\nabla u(x,t) + Ru(x,t) \qquad (IV.2)$$

With D is the hydrodynamic dispersion tensor, V is the flow velocity field and R is a

Chapitre IV. Inverse source problem based on two dimensionless dispersion-current functions in 2D evolution transport equations with spatially varying coefficients

real number that represents the first order decay reaction coefficient. Moreover, the velocity $V = (V_1, V_2)^\top$ is a spatially varying field that satisfies

$$\text{rot}(V) = \text{div}(V) = 0, \quad V_1 > 0, \quad V_2 \geq 0 \text{ a.e. in } \Omega \quad \text{and} \quad V.\nu = 0 \quad \text{on } \Gamma_L \qquad \text{(IV.3)}$$

Here, ν is the exterior unit normal vector to Ω on $\partial\Omega$. Hydrodynamic dispersion occurs as a consequence of two processes : molecular diffusion resulting from the random molecular motion and mechanical dispersion which is caused by non-uniform velocities. The summation of these two processes defines the hydrodynamic dispersion tensor [5] :

$$D = D_M \mathbf{I} + \begin{bmatrix} D_1 & D_0 \\ D_0 & D_2 \end{bmatrix} \qquad \text{(IV.4)}$$

With $D_M > 0$ is a real number that represents the molecular diffusion coefficient, $\mathbf{I}$ is the 2×2 identity matrix and the spatially varying entries $D_{i=0,1,2}$ are such that, see [48, 49] :

$$D_0 = \frac{V_1 V_2 (D_L - D_T)}{\|V\|_2^2}, \quad D_1 = \frac{D_L V_1^2 + D_T V_2^2}{\|V\|_2^2} \quad \text{and} \quad D_2 = \frac{D_L V_2^2 + D_T V_1^2}{\|V\|_2^2} \qquad \text{(IV.5)}$$

Where D_T and D_L are the transverse and longitudinal dispersion coefficients that satisfy $0 \leq D_T < D_L$. Thus, according to (IV.4)-(IV.5) the dispersion tensor D can be rewritten as

$$D = (D_M + D_T)\mathbf{I} + \frac{D_L - D_T}{\|V\|_2^2} V V^\top \implies (D_M + D_T)\|X\|_2^2 \leq DX.X \leq (D_M + D_L)\|X\|_2^2 \qquad \text{(IV.6)}$$

For all $X \in \mathbb{R}^2$. Hence, (IV.6) implies that the matrix D is uniformly elliptic and bounded in Ω. In the remainder, we assume V_1, V_2 and D_L, D_T to be Lipschitz functions in Ω.

Besides, to (IV.1)-(IV.2) one has to append initial and boundary conditions. Moreover, we could use without loss of generality no pollution occurring at the initial monitoring time and thus, a null initial BOD concentration. As far as the boundary conditions are concerned, an homogeneous Dirichlet condition on the inflow boundary seems to be reasonable since the convective transport generally dominates the diffusion process. However, other physical considerations suggest to use rather a Neumann homogeneous condition on the remaining parts of

IV.2 Problem statement

the boundary. Then, we employ the following conditions :

$$\begin{aligned} u(.,0) &= 0 & &\text{in } \Omega \\ u &= 0 & &\text{on } \Sigma_D := \Gamma_D \times (0,T) \\ D\nabla u.\nu &= 0 & &\text{on } \Sigma_N := \Gamma_N \times (0,T) \end{aligned} \qquad (IV.7)$$

Notice that due to the linearity of the operator L introduced in (IV.2) and according to the superposition principle, the use of a non-zero initial condition and/or inhomogeneous boundary conditions do not affect the results established in this paper.

As far as F is concerned, we consider a time-dependent point source defined by

$$F(x,t) = \lambda(t)\delta(x-S) \qquad \text{for all } x = (x_1,x_2) \in \Omega \text{ and } t \in (0,T) \qquad (IV.8)$$

Where δ denotes the Dirac mass, $S = (S_{x_1}, S_{x_2})$ is an interior point of Ω that represents the source position and $\lambda(.) \in L^2(0,T)$ designates its time-dependent intensity function. Then, the transposition method introduced by Lions [47] implies that for a given F as in (IV.8), the problem introduced in (IV.1)-(IV.7) admits a unique solution u which belongs to

$$L^2\big(0,T;L^2(\Omega)\big) \cap \mathcal{C}^0\big(0,T;H^{-1}(\Omega)\big) \qquad (IV.9)$$

Moreover, since the source position S is assumed to be an interior point of Ω, the state u is smooth enough on $\partial\Omega$ which allows to define the boundary observation operator :

$$M[F] := \Big\{ D\nabla u.\nu \text{ on } \Sigma_D, \ (V^\perp.\nu)u \text{ on } \Sigma_L, \ u \text{ on } \Sigma_{out} \Big\} \qquad (IV.10)$$

With $V^\perp.\nu$ is the tangential component of the velocity field V on the solid boundary $\Sigma_L := \Gamma_L \times (0,T)$ whereas $\Sigma_{out} := \Gamma_{out} \times (0,T)$. This is the so-called *forward problem*.

The *inverse problem* with which we are concerned here is : given d_{in}, d_L and d_{out} boundary records of $D\nabla u.\nu$ on Σ_D and of u on Σ_L, Σ_{out} determine the position S and the time-dependent intensity $\lambda(.)$ defining a sought source F as introduced in (IV.8) that yields

Chapitre IV. Inverse source problem based on two dimensionless dispersion-current functions in 2D evolution transport equations with spatially varying coefficients

$$M[F] = \left\{ d_{in} \text{ on } \Sigma_D, \ (V^\perp.\nu)d_L \text{ on } \Sigma_L, \ d_{out} \text{ on } \Sigma_{out} \right\} \tag{IV.11}$$

Notice that if the velocity field V satisfies on the solid boundary Γ_L the so-called *no-slipping* condition i.e., $V = \vec{0}$ on Γ_L then, the identification results established in this paper do not require the records of the state u on Γ_L. Besides, regarding the regularity of u at $t \in (0,T)$, the authors in [16, 58] proved for a similar problem that the assumption :

$$\exists T^0 \in (0,T) \quad \text{such that} \quad \lambda(t) = 0 \quad \text{for all } t \in (T^0, T) \tag{IV.12}$$

Leads the solution u of the problem (IV.1)-(IV.7) with a source F as in (IV.8)-(IV.12) to satisfy

$$u(.,T^*) \in L^2(\Omega) \quad \text{for all } T^* \in (T^0, T) \tag{IV.13}$$

For later use and as in view of (IV.6) the matrix D is invertible in Ω then, there exists a unique vector field X solution of the linear system $DX + V = 0$ in Ω. Moreover, we have

$$D^{-1} = \frac{1}{\det(D)} \begin{bmatrix} D_M + D_2 & -D_0 \\ -D_0 & D_M + D_1 \end{bmatrix} \implies X = -\frac{1}{\det(D)} \begin{bmatrix} D_M V_1 + D_2 V_1 - D_0 V_2 \\ D_M V_2 + D_1 V_2 - D_0 V_1 \end{bmatrix} \tag{IV.14}$$

Furthermore, according to (IV.5) we find

$$\begin{aligned} \det(D) &= \big(D_M + D_L\big)\big(D_M + D_T\big) \\ D_2 V_1 - D_0 V_2 &= V_1 D_T \\ D_1 V_2 - D_0 V_1 &= V_2 D_T \end{aligned} \tag{IV.15}$$

Thus, using the results (IV.15) to substitute the associated terms in the right-hand side of the second equation in (IV.14) gives $X = -\frac{1}{D_M + D_L} V$. Then, provided the following condition :

IV.2 Problem statement

$$\text{rot}\left(\frac{V}{D_M + D_L}\right) = 0 \tag{IV.16}$$

Holds true in Ω, X is a gradient field derived from a scalar potential ψ that solves

$$D\nabla\psi + V = 0 \quad \Leftrightarrow \quad \nabla\psi = -\frac{1}{D_M + D_L}V \tag{IV.17}$$

Hence, ψ that satisfies $\psi(a,b) = 0$ with $(a,b) \in \Omega$ is defined as follows :

$$\psi(x_1, x_2) = -\int_a^{x_1}\left(\frac{V_1}{D_M + D_L}\right)(\eta, x_2)d\eta - \int_b^{x_2}\left(\frac{V_2}{D_M + D_L}\right)(a, \zeta)d\zeta \tag{IV.18}$$

Likewise, using the vector $V^\perp = (-V_2, V_1)^\top$ we consider the following equation :

$$D\nabla\psi^\perp + V^\perp = 0 \quad \Leftrightarrow \quad \nabla\psi^\perp = -\frac{1}{D_M + D_T}V^\perp \tag{IV.19}$$

The second equation in (IV.19) is obtained using similar techniques as in (IV.14)-(IV.15). Thus, provided the following condition :

$$\text{div}\left(\frac{V}{D_M + D_T}\right) = 0 \tag{IV.20}$$

Holds true in Ω, the scalar potential $\psi^\perp$ that solves (IV.19) with $\psi^\perp(a,b) = 0$ is defined by

$$\psi^\perp(x_1, x_2) = \int_a^{x_1}\left(\frac{V_2}{D_M + D_T}\right)(\eta, x_2)d\eta - \int_b^{x_2}\left(\frac{V_1}{D_M + D_T}\right)(a, \zeta)d\zeta \tag{IV.21}$$

Therefore, in view of (IV.3) and under the conditions (IV.16)-(IV.20), the three functions

$$z_g(x,t) = e^{g(x) - R(T-t)} \quad \text{for} \quad g = 0, \psi \quad \text{and} \quad z^\perp(x,t) = \psi^\perp(x)e^{-R(T-t)} \tag{IV.22}$$

Solve the following adjoint equation :

Chapitre IV. Inverse source problem based on two dimensionless dispersion-current functions in 2D evolution transport equations with spatially varying coefficients

$$-\partial_t z - \text{div}\left(D\nabla z\right) - V\nabla z + Rz = 0 \quad \text{in } \Omega \times (0,T) \tag{IV.23}$$

Remark 2.1 *Note that in view of (IV.17) and (IV.19), the two dimensionless dispersion-current functions ψ and $\psi^\perp$ are of orthogonal gradients in Ω. This property is an essential ingredient of the identifiability theorem established later in the present study since it means that each intersection between two level sets of ψ and $\psi^\perp$ happens always at a single point.*

Then, we prove the following lemma that establishes a link between the coordinates of two points lying on a same level set of ψ and $\psi^\perp$:

Lemma 2.2 *Let $\hat{x} = (\hat{x}_1, \hat{x}_2)$ and $\tilde{x} = (\tilde{x}_1, \tilde{x}_2)$ be two interior points of Ω. Provided the two conditions (IV.16) and (IV.20) hold true in Ω, we have*

$$\begin{aligned}\psi(\hat{x}) = \psi(\tilde{x}) &\Leftrightarrow \int_{\hat{x}_1}^{\tilde{x}_1}\left(\frac{V_1}{D_M + D_L}\right)(\eta, \tilde{x}_2)d\eta = -\int_{\hat{x}_2}^{\tilde{x}_2}\left(\frac{V_2}{D_M + D_L}\right)(\hat{x}_1, \zeta)d\zeta \\ \psi^\perp(\hat{x}) = \psi^\perp(\tilde{x}) &\Leftrightarrow \int_{\hat{x}_1}^{\tilde{x}_1}\left(\frac{V_2}{D_M + D_T}\right)(\eta, \tilde{x}_2)d\eta = \int_{\hat{x}_2}^{\tilde{x}_2}\left(\frac{V_1}{D_M + D_T}\right)(\hat{x}_1, \zeta)d\zeta\end{aligned} \tag{IV.24}$$

Where ψ and $\psi^\perp$ are the two dispersion-current functions obtained in (IV.18) and (IV.21).

Proof. Let $\hat{x} = (\hat{x}_1, \hat{x}_2)$ and $\tilde{x} = (\tilde{x}_1, \tilde{x}_2)$ be two interior points of Ω. From (IV.18), by setting $\psi(\hat{x}) = \psi(\tilde{x})$ and using Chasles's rule on the integral over $(a, \tilde{x}_1)$, we obtain

$$\int_{\hat{x}_2}^{\tilde{x}_2}\left(\frac{V_2}{D_M + D_L}\right)(a, \zeta)d\zeta = -\int_{\hat{x}_1}^{\tilde{x}_1}\left(\frac{V_1}{D_M + D_L}\right)(\eta, \tilde{x}_2)d\eta + \int_a^{\hat{x}_1}\left[\left(\frac{V_1}{D_M + D_L}\right)(\eta, \zeta)\right]_{\zeta=\tilde{x}_2}^{\zeta=\hat{x}_2}d\eta \tag{IV.25}$$

Moreover, in view of (IV.16) and using Fubini's rule we find

$$\begin{aligned}\int_a^{\hat{x}_1}\left[\left(\frac{V_1}{D_M + D_L}\right)(\eta, \zeta)\right]_{\zeta=\tilde{x}_2}^{\zeta=\hat{x}_2}d\eta &= \int_a^{\hat{x}_1}\left(\int_{\tilde{x}_2}^{\hat{x}_2}\partial_\zeta\left(\frac{V_1}{D_M + D_L}\right)(\eta, \zeta)d\zeta\right)d\eta \\ &= \int_{\tilde{x}_2}^{\hat{x}_2}\left(\int_a^{\hat{x}_1}\partial_\eta\left(\frac{V_2}{D_M + D_L}\right)(\eta, \zeta)d\eta\right)d\zeta \\ &= -\int_{\hat{x}_2}^{\tilde{x}_2}\left[\left(\frac{V_2}{D_M + D_L}\right)(\eta, \zeta)\right]_{\eta=a}^{\eta=\hat{x}_1}d\zeta\end{aligned} \tag{IV.26}$$

IV.2 Problem statement

Thus, substituting the last term in the equality (IV.25) by its value obtained in (IV.26) gives the first result announced in (IV.51). Besides, in view of (IV.21), by setting $\psi^\perp(\hat{x}) = \psi^\perp(\tilde{x})$ and using Chasles's rule on the integral over $(a, \tilde{x}_1)$, we get

$$\int_{\hat{x}_2}^{\tilde{x}_2} \left(\frac{V_1}{D_M + D_T} \right)(a, \zeta) d\zeta = \int_{\hat{x}_1}^{\tilde{x}_1} \left(\frac{V_2}{D_M + D_T} \right)(\eta, \tilde{x}_2) d\eta + \int_{a}^{\hat{x}_1} \left[\left(\frac{V_2}{D_M + D_T} \right)(\eta, \zeta) \right]_{\zeta=\hat{x}_2}^{\zeta=\tilde{x}_2} d\eta \quad \text{(IV.27)}$$

Furthermore, in view of (IV.20) and using Fubini's rule we find

$$\begin{aligned}
\int_{a}^{\hat{x}_1} \left[\left(\frac{V_2}{D_M + D_T} \right)(\eta, \zeta) \right]_{\zeta=\hat{x}_2}^{\zeta=\tilde{x}_2} d\eta &= \int_{a}^{\hat{x}_1} \left(\int_{\hat{x}_2}^{\tilde{x}_2} \partial_\zeta \left(\frac{V_2}{D_M + D_T} \right)(\eta, \zeta) d\zeta \right) d\eta \\
&= -\int_{\hat{x}_2}^{\tilde{x}_2} \left(\int_{a}^{\hat{x}_1} \partial_\eta \left(\frac{V_1}{D_M + D_T} \right)(\eta, \zeta) d\eta \right) d\zeta \quad \text{(IV.28)} \\
&= -\int_{\hat{x}_2}^{\tilde{x}_2} \left[\left(\frac{V_1}{D_M + D_T} \right)(\eta, \zeta) \right]_{\eta=a}^{\eta=\hat{x}_1} d\zeta
\end{aligned}$$

From substituting the last term in the right hand-side of (IV.27) by its value found in (IV.28), we obtain the second result announced in (IV.51). □

Besides, we introduce the following boundary null controllability problem : given $\varphi_0 \in L^2(\Omega)$, determine a boundary control $\gamma \in L^2\big(\Gamma_{out} \times (T^0, T)\big)$ that drives the solution φ of

$$\begin{cases}
\partial_t \varphi - \text{div}(D\nabla\varphi) - V\nabla\varphi + R\varphi = 0 & \text{in } \Omega \times (T^0, T) \\
\varphi(., T^0) = \varphi_0 & \text{in } \Omega \\
D\nabla\varphi.\nu = 0 & \text{on } (\Gamma_D \cup \Gamma_L) \times (T^0, T) \\
\varphi = \gamma & \text{on } \Gamma_{out} \times (T^0, T)
\end{cases} \quad \text{(IV.29)}$$

$$\text{To satisfy : } \varphi(., T) = 0 \quad \text{in } \Omega \quad \text{(IV.30)}$$

Let $\Phi(x, t) = e^{-\frac{1}{2}\psi(x)} \varphi(x, t)$ in $\Omega \times (T^0, T)$ where ψ is the function defined in (IV.18). Then, the boundary null controllability problem introduced in (IV.29)-(IV.30) is equivalent to find $\gamma \in L^2\big(\Gamma_{out} \times (T^0, T)\big)$ such that the solution Φ of the system :

Chapitre IV. Inverse source problem based on two dimensionless dispersion-current functions in 2D evolution transport equations with spatially varying coefficients

$$\begin{cases} \partial_t \Phi - \text{div}(D\nabla\Phi) + \rho\Phi = 0 & \text{in } \Omega \times (T^0, T) \\ \Phi(.,T^0) = e^{-\frac{1}{2}\psi}\varphi_0 & \text{in } \Omega \\ \left(D\nabla\Phi - \frac{1}{2}\Phi V\right).\nu = 0 & \text{on } \Gamma_D \times (T^0, T) \\ D\nabla\Phi.\nu = 0 & \text{on } \Gamma_L \times (T^0, T) \\ \Phi = e^{-\frac{1}{2}\psi}\gamma & \text{on } \Gamma_{out} \times (T^0, T) \end{cases} \quad (IV.31)$$

$$\text{satisfies} \quad \Phi(.,T) = 0 \quad \text{in } \Omega \qquad (IV.32)$$

Where $\rho = R + \frac{1}{4}V^\top D^{-1} V = R + \|V\|_2^2 / 4(D_M + D_L)$. Furthermore, we introduce the adjoint problem associated to (IV.31) that is to a given $\xi_0 \in L^2(\Omega)$, determine ξ that solves

$$\begin{cases} -\partial_t \xi - \text{div}(D\nabla\xi) + \rho\xi = 0 & \text{in } \Omega \times (T^0, T) \\ \xi(.,T) = \xi_0 & \text{in } \Omega \\ \left(D\nabla\xi - \frac{1}{2}\xi V\right).\nu = 0 & \text{on } \Gamma_D \times (T^0, T) \\ D\nabla\xi.\nu = 0 & \text{on } \Gamma_L \times (T^0, T) \\ \xi = 0 & \text{on } \Gamma_{out} \times (T^0, T) \end{cases} \quad (IV.33)$$

In addition, let $J : L^2(\Omega) \to \mathbb{R}$ be the functional defined for a given ξ_0 as follows :

$$J(\xi_0) = \frac{1}{2} \prec D\nabla\xi.\nu, D\nabla\xi.\nu \succ_{L^2(\Gamma_{out} \times (T^0,T))} - \prec e^{-\frac{1}{2}\psi}\varphi_0, \xi(.,T^0) \succ_{L^2(\Omega)} \qquad (IV.34)$$

Then, we establish the following theorem :

Theorem 2.3 (Boundary null controllability - *HUM* control)

Provided V_1, V_2 and D_L, D_T are Lipschitz functions that satisfy (IV.16) then, for $T^0 \in (0, T)$ and given $\varphi_0 \in L^2(\Omega)$ the boundary control

$$e^{-\frac{1}{2}\psi}\gamma = D\nabla\hat{\xi}.\nu \quad \text{on} \quad \Gamma_{out} \times (T^0, T) \qquad (IV.35)$$

Is the control of smallest $L^2\big(\Gamma_{out} \times (T^0, T)\big)$-norm that leads the solution Φ of (IV.31) to satisfy (IV.32). Here, ψ is the function defined in (IV.18) and $\hat{\xi}$ is the solution of (IV.33) with

IV.2 Problem statement

$\hat{\xi}(.,T) = \hat{\xi}_0$ where $\hat{\xi}_0$ is the unique minimizer of the functional J introduced in (IV.34).

Proof. To establish the proof of Theorem 2.1, we proceed in two steps :

Step 1. We start by proving that a control $\gamma \in L^2\big(\Gamma_{out} \times (T^0,T)\big)$ solves the boundary null controllability problem (IV.31)-(IV.32) if and only if

$$\prec e^{-\frac{1}{2}\psi}\varphi_0, \xi(.,T^0) \succ_{L^2(\Omega)} = \prec e^{-\frac{1}{2}\psi}\gamma, D\nabla\xi.\nu \succ_{L^2(\Gamma_{out}\times(T^0,T))}, \quad \forall \xi_0 \in L^2(\Omega) \qquad \text{(IV.36)}$$

ξ is the solution of the adjoint problem (IV.33). Indeed, using Green's formula with the two problems (IV.31) and (IV.33), it comes that this first step is an immediate consequence of

$$\left[\prec \Phi, \xi \succ_{L^2(\Omega)}\right]_{T^0}^T = \int_{T^0}^T \left(\prec \partial_t\Phi, \xi \succ_{L^2(\Omega)} + \prec \Phi, \partial_t\xi \succ_{L^2(\Omega)}\right) dt \qquad \text{(IV.37)}$$
$$= -\prec \Phi, D\nabla\xi.\nu \succ_{L^2(\Gamma_{out}\times(T^0,T))}$$

Since $\Phi(.,T) = 0$ in Ω, the sufficient condition is a straightforward consequence of (IV.37). As far as the necessary condition is concerned, assuming $\gamma \in L^2\big(\Gamma_{out} \times (T^0,T)\big)$ leads the solution Φ of (IV.31) to satisfy (IV.36) then, by identification with (IV.37) we find

$$\prec \Phi(.,T), \xi_0 \succ_{L^2(\Omega)} = 0 \quad \text{for all} \quad \xi_0 \in L^2(\Omega)$$

Step 2. Provided there exists $\hat{\xi}_0 \in L^2(\Omega)$ a minimizer of the functional J, we have

$$\nabla J(\hat{\xi}_0).\xi_0 = \prec D\nabla\hat{\xi}.\nu, D\nabla\xi.\nu \succ_{L^2(\Gamma_{out}\times(T^0,T))} - \prec e^{-\frac{1}{2}\psi}\varphi_0, \xi(.,T^0) \succ_{L^2(\Omega)} \qquad \text{(IV.38)}$$
$$= 0 \quad \text{for all } \xi_0 \in L^2(\Omega)$$

Where $\hat{\xi}$ and ξ are the solutions of the adjoint problem (IV.33) with the initial data $\hat{\xi}(.,T) = \hat{\xi}_0$ and $\xi(.,T) = \xi_0$. Hence, in view of (IV.38), the boundary control γ defined from

$$e^{-\frac{1}{2}\psi}\gamma = D\nabla\hat{\xi}.\nu \qquad \text{on } \Gamma_{out} \times (T^0,T) \qquad \text{(IV.39)}$$

Fulfills (IV.36) and thus, it solves the boundary null controllability problem (IV.31)-(IV.32). Furthermore, suppose that $e^{-\frac{1}{2}\psi}\tilde{\gamma} \in L^2\big(\Gamma_{out} \times (T^0,T)\big)$ is also another boundary control that

Chapitre IV. Inverse source problem based on two dimensionless dispersion-current functions in 2D evolution transport equations with spatially varying coefficients

leads the solution Φ of (IV.31) to satisfy (IV.32). Then, in view of (IV.36), we have for all $\xi_0 \in L^2(\Omega)$ and ξ solution of the adjoint problem (IV.33) with $\xi(.,T) = \xi_0$ that

$$\prec e^{-\frac{1}{2}\psi}\gamma, D\nabla\xi.\nu \succ_{L^2(\Gamma_{out}\times(T^0,T))} = \prec e^{-\frac{1}{2}\psi}\tilde{\gamma}, D\nabla\xi.\nu \succ_{L^2(\Gamma_{out}\times(T^0,T))} \qquad \text{(IV.40)}$$

Using $\xi = \hat{\xi}$ in (IV.40) and in view of (IV.39), we find $\|e^{-\frac{1}{2}\psi}\gamma\|_{L^2(\Gamma_{out}\times(T^0,T))} \leq \|e^{-\frac{1}{2}\psi}\tilde{\gamma}\|_{L^2(\Gamma_{out}\times(T^0,T))}$.

□

Proposition 2.4 *Under the assumptions of Theorem 2.1, the functional J introduced in (IV.34) admits a unique minimizer.*

Proof. See the appendix.

Remark 2.5 *The control γ defined from (IV.35) is the so-called HUM boundary control with refering to the Hilbert Uniqueness Method introduced by J. Lions [45, 46].*

Here, we prove the last technical result needed to establish our identifiability theorem.

Lemma 2.6 *Provided V_1, V_2 and D_L, D_T are Lipschitz functions, for all $x^* \in \Omega$ there exists $\zeta_{in} \in L^2(\Gamma_D \times (0, T^0))$ and $\zeta_{out} \in L^2(\Gamma_{out} \times (0, T^0))$ that maintain the solution of*

$$\begin{cases} \partial_t v - \text{div}(D\nabla v) - V\nabla v + Rv = 0 & \text{in } \Omega \times (0, T^0) \\ v(.,0) = 0 & \text{in } \Omega \\ v = \zeta_{in} & \text{on } \Gamma_D \times (0, T^0) \\ D\nabla v.\nu = V^\perp.\nu & \text{on } \Gamma_L \times (0, T^0) \\ (D\nabla v + vV).\nu = \zeta_{out} & \text{on } \Gamma_{out} \times (0, T^0) \end{cases} \qquad \text{(IV.41)}$$

$$\text{Such that} \quad v(x^*,.) \neq 0 \quad \text{almost everywhere in } (0, T^0) \qquad \text{(IV.42)}$$

Proof. Let x^* be an interior point of Ω and θ be the solution of the adjoint problem :

IV.2 Problem statement

$$\begin{aligned}
-\partial_t \theta - \operatorname{div}(D\nabla\theta) + V\nabla\theta + R\theta &= \delta(x-x^*) \quad \text{in } \Omega \times (0, T^0) \\
\theta(., T^0) &= 0 \quad \text{in } \Omega \\
\theta &= 0 \quad \text{on } \Gamma_D \times (0, T^0) \\
D\nabla\theta.\nu &= 0 \quad \text{on } \Gamma_N \times (0, T^0)
\end{aligned} \quad \text{(IV.43)}$$

Then, multiplying the first equation in (IV.43) by the solution v of the problem (IV.41) and integrating by parts over Ω using Green's formula gives for all $t \in (0, T^0)$

$$v(x^*, t) = -\frac{d}{dt} \prec \theta(., t), v(., t) \succ + \int_{\Gamma_{out}} \theta \zeta_{out} d\Gamma - \int_{\Gamma_D} \zeta_{in} D\nabla\theta.\nu d\Gamma + \int_{\Gamma_L} \theta V^{\perp}.\nu d\Gamma \quad \text{(IV.44)}$$

Suppose that for all ζ_{in} and ζ_{out} there exists $0 \leq t_0 < t_1 \leq T^0$ such that $v(x^*, .) = 0$ in (t_0, t_1). As according to [10] the time-dependent function $v(x^*, .)$ is analytic in $(0, T^0)$, it follows that $v(x^*, .) = 0$ in $(0, T^0)$. Therefore, by integrating (IV.44) over $(0, T^0)$ we obtain

$$\int_{\Gamma_L \times (0, T^0)} \theta V^{\perp}.\nu d\Gamma dt + \int_{\Gamma_{out} \times (0, T^0)} \theta \zeta_{out} d\Gamma dt - \int_{\Gamma_D \times (0, T^0)} \zeta_{in} D\nabla\theta.\nu d\Gamma dt = 0, \quad \forall \zeta_{in}, \zeta_{out}$$

Thus, $\theta = 0$ on $\Gamma_{out} \times (0, T^0)$ and $D\nabla\theta.\nu = 0$ on $\Gamma_D \times (0, T^0)$. That implies in view of the unique continuation Theorem from [42], $\theta = 0$ in $(\Omega \setminus \{x^*\}) \times (0, T^0)$ which is absurd since using the transposition method [47], we have $\theta \in L^2(\Omega \times (0, T^0))$. Therefore, there exists always two boundary controls $\zeta_{in} \in L^2(\Gamma_D \times (0, T^0))$ and $\zeta_{out} \in L^2(\Gamma_{out} \times (0, T^0))$ that maintain the associated solution of the problem (IV.41) such that (IV.42) holds true. □

Remark 2.7 *Due to the analyticity of the time-dependent function $v(x^*, .)$ in $(0, T^0)$ see [10], the unique continuation Theorem from [42] and according to (IV.44), we can ensure the assertion (IV.42) by employing in the problem (IV.41) the boundary controls ζ_{in} and ζ_{out} such that : If $\int_{\Gamma_L \times (0, T^0)} \theta V^{\perp}.\nu d\Gamma dt \geq 0$ then, we use either $\zeta_{out} = \theta$ and $\zeta_{in} = 0$ or $\zeta_{out} = 0$ and $\zeta_{in} = -D\nabla\theta.\nu$ or $\zeta_{out} = \theta$ and $\zeta_{in} = -D\nabla\theta.\nu$ Otherwise, we use either $\zeta_{out} = -\theta$ and $\zeta_{in} = 0$ or $\zeta_{out} = 0$ and $\zeta_{in} = D\nabla\theta.\nu$ or $\zeta_{out} = -\theta$ and $\zeta_{in} = D\nabla\theta.\nu$ Therefore, we can always ensure (IV.42) using an only one control. In practice, this choice can be motivated by the nature of the flow (dominant advection, dominant diffusion) and/or the position of x^* with respect to the inflow/outflow boundary.*

Chapitre IV. Inverse source problem based on two dimensionless dispersion-current functions in 2D evolution transport equations with spatially varying coefficients

3 Identifiability

In this section, we prove that under some reasonable assumptions the boundary observation operator $M[F]$ introduced in (IV.10) enables to uniquely identify the elements S and $\lambda(.)$ defining the sought time-dependent point source F presented in (IV.8). To this end, we establish the following identifiability Theorem :

Theorem 3.1 *Let $F^{(i)}(x,t) = \lambda^{(i)}(t)\delta(x - S^{(i)})$ where $\lambda^{(i)}(.) \in L^2(0,T)$ that satisfies (IV.12), $S^{(i)} = (S^{(i)}_{x_1}, S^{(i)}_{x_2})$ is an interior point of Ω and $M[F^{(i)}]$ be the boundary observation operator introduced in (IV.10) and associated to the source $F^{(i)}$, for $i = 1, 2$. Provided $V = (V_1, V_2)^\top$ satisfies (IV.3) with V_1, V_2 and D_L, D_T are Lipschitz functions such that (IV.16) and (IV.20) hold true in Ω, we have*

$$M[F^{(1)}] = M[F^{(2)}] \implies S^{(1)} = S^{(2)} \text{ and } \lambda^{(1)}(.) = \lambda^{(2)}(.) \quad a.e. \text{ in } (0,T) \quad \text{(IV.45)}$$

Proof. Let $u_{i=1,2}$ be the solution of the problem (IV.1)-(IV.7) with $F = F^{(i)}$, for $i = 1, 2$ and $w = u_2 - u_1$. Then, the variable w solves the following system :

$$\begin{aligned}
L[w](x,t) &= \lambda^{(2)}(t)\delta(x - S^{(2)}) - \lambda^{(1)}(t)\delta(x - S^{(1)}) && \text{in } \Omega \times (0,T) \\
w(.,0) &= 0 && \text{in } \Omega \\
w &= 0 && \text{on } \Gamma_D \times (0,T) \\
D\nabla w.\nu &= 0 && \text{on } \Gamma_N \times (0,T)
\end{aligned} \quad \text{(IV.46)}$$

Furthermore, according to (IV.10), having $M[F^{(1)}] = M[F^{(2)}]$ implies that

$$D\nabla w.\nu = 0 \text{ on } \Sigma_D, \quad (V^\perp.\nu)w = 0 \text{ on } \Sigma_L \quad \text{and} \quad w = 0 \text{ on } \Sigma_{out} \quad \text{(IV.47)}$$

Hence, multiplying the first equation in (IV.46) by z that satisfies the adjoint equation (IV.23) and integrating by parts over Ω gives using Green's formula and (IV.47) that for all $t \in (0,T)$

IV.3 Identifiability

$$\lambda^{(2)}(t)z(S^{(2)},t) - \lambda^{(1)}(t)z(S^{(1)},t) = \frac{d}{dt} \prec w(.,t), z(.,t) \succ + \int_{\Gamma_L} wD\nabla z.\nu d\Gamma \qquad \text{(IV.48)}$$

In addition, since $\lambda^{(1)}(.)$ and $\lambda^{(2)}(.)$ fulfill the assumption (IV.12) with $T^0 \in (0,T)$, the variable w satisfies in $\Omega \times (T^*, T)$ with $T^* \in (T^0, T)$ a system similar to (IV.46) where the first equation becomes homogeneous and the initial data is $w(.,T^*) \in L^2(\Omega)$. Therefore, in view of (IV.47) and using the unique continuation Theorem from [42] we obtain $w(.,T^*) = 0$ a.e. in Ω. Thus, by integrating (IV.48) over $(0,T^*)$ we find

$$\int_0^{T^0} \Big(\lambda^{(2)}(t)z(S^{(2)},t) - \lambda^{(1)}(t)z(S^{(1)},t)\Big) dt = \int_{\Gamma_L \times (0,T^*)} wD\nabla z.\nu d\Gamma dt \qquad \text{(IV.49)}$$

The three functions $z_g(x,t) = e^{g(x)-R(T-t)}$ for $g = 0, \psi$ and $z^\perp(x,t) = \psi^\perp(x)e^{-R(T-t)}$ introduced in (IV.22) solve the adjoint equation (IV.23) and, according to (IV.17)-(IV.19), satisfy on $\Gamma_L \times (0,T) : D\nabla z_g.\nu = 0$ and $D\nabla z^\perp.\nu = -e^{-R(T-t)}V^\perp.\nu$ Therefore, in view of (IV.47), the right-hand side in (IV.49) vanishes for the three functions : $z = z_0$, $z = z_\psi$ and $z = z^\perp$. Then, from (IV.49) it comes that the two source positions $S^{(1)}$ and $S^{(2)}$ solve the following system :

$$\begin{aligned} \psi(S^{(1)}) &= \psi(S^{(2)}) \\ \psi^\perp(S^{(1)}) &= \psi^\perp(S^{(2)}) \end{aligned} \qquad \text{(IV.50)}$$

Furthermore, according to Lemma 2.2, the system (IV.50) is equivalent to

$$\begin{aligned} \int_{S_{x_1}^{(1)}}^{S_{x_1}^{(2)}} \left(\frac{V_1}{D_M + D_L}\right)(\eta, S_{x_2}^{(2)}) d\eta &= -\int_{S_{x_2}^{(1)}}^{S_{x_2}^{(2)}} \left(\frac{V_2}{D_M + D_L}\right)(S_{x_1}^{(1)}, \zeta) d\zeta \\ \int_{S_{x_1}^{(1)}}^{S_{x_1}^{(2)}} \left(\frac{V_2}{D_M + D_T}\right)(\eta, S_{x_2}^{(2)}) d\eta &= \int_{S_{x_2}^{(1)}}^{S_{x_2}^{(2)}} \left(\frac{V_1}{D_M + D_T}\right)(S_{x_1}^{(1)}, \zeta) d\zeta \end{aligned} \qquad \text{(IV.51)}$$

Since $V_1 > 0$ and $V_2 \geq 0$ a.e. in Ω then, in the system (IV.51) we have at least one of the four involved integrals is null either because $V_2 = 0$ or because the two left-hand side terms have a same sign whereas the two right-hand side terms have aposite signs. That implies $S_{x_1}^{(1)} = S_{x_1}^{(2)}$ and $S_{x_2}^{(1)} = S_{x_2}^{(2)}$ i.e., $S^{(1)} = S^{(2)}$.

Chapitre IV. Inverse source problem based on two dimensionless dispersion-current functions in 2D evolution transport equations with spatially varying coefficients

We set $S^{(1)} = S^{(2)} = S$ in (IV.46). Let ζ_{in} and ζ_{out} be such that the solution v of the problem (IV.41) yields the assertion (IV.42) with $x^* = S$ i.e., $v(S,.) \neq 0$ a.e. in $(0, T^0)$. Then, using v and given $\tau \in (0, T^0)$ we define the variable $\tilde{v}_\tau(.,t) = v(.,\tau-t)$ for all $t \in (0,\tau)$. Thus, $\tilde{v}_\tau$ solves the following system :

$$\begin{cases} -\partial_t \tilde{v}_\tau - \mathrm{div}(D\nabla \tilde{v}_\tau) - V\nabla \tilde{v}_\tau + R\tilde{v}_\tau = 0 & \text{in } \Omega \times (0,\tau) \\ \tilde{v}_\tau(.,\tau) = 0 & \text{in } \Omega \\ \tilde{v}_\tau(.,t) = \zeta_{in}(.,\tau-t) & \text{on } \Gamma_D \times (0,\tau) \\ D\nabla \tilde{v}_\tau.\nu = V^\perp.\nu & \text{on } \Gamma_L \times (0,\tau) \\ \left(D\nabla \tilde{v}_\tau(.,t) + \tilde{v}_\tau(.,t)V\right).\nu = \zeta_{out}(.,\tau-t) & \text{on } \Gamma_{out} \times (0,\tau) \end{cases} \quad (\text{IV.52})$$

Furthermore, multiplying the first equation in (IV.46) by $\tilde{v}_\tau$ and integrating by parts over Ω using the boundary conditions given in (IV.46)-(IV.47) then, integrating the obtained result with respect to the time over $(0,\tau)$ leads to

$$\int_0^\tau \left(\lambda^{(2)}(t) - \lambda^{(1)}(t)\right) v(S, \tau - t) dt = 0, \qquad \text{for all } \tau \in (0, T^0) \quad (\text{IV.53})$$

Using Titchmarsh's theorem on convolution of L^1 functions [60] and since $v(S,.) \neq 0$ a.e. in $(0, T^0)$ it follows from (IV.53) that $\lambda^{(1)}(.) = \lambda^{(2)}(.)$ a.e. in $(0, T^0)$. □

4 Identification

Given the boundary records introduced in (IV.11) related to the state u solution of the problem (IV.1)-(IV.7) with an unknown source F as defined in (IV.8), we focus in this section on establishing an identification method that uses those records to : localize the source position S, give lower and upper frame bounds of the total amount loaded by the time-dependent source intensity function $\lambda(.)$ and identify the historic of this latest. To this end, we proceed in two steps : Firstly, we localize the sought position S as the unique solution of a nonlinear system of two equations made by the two dispersion-current functions ψ, $\psi^\perp$ and determine the two frame bounds. Secondly, we transform the task of identifying $\lambda(.)$ into solving a deconvolution problem.

IV.4 Identification

4.1 Localization of the sought source position

Proposition 4.1 *Let u be the solution of the problem (IV.1)-(IV.7) with an unknown source $F(x,t) = \lambda(t)\delta(x-S)$ where $\lambda(.) \in L^2(0,T)$ that satisfies (IV.12) and $M[F]$ be the associated boundary observation operator introduced in (IV.10). Provided the conditions (IV.3), (IV.16) and (IV.20) hold true then, $M[F] = \{d_{in} \text{ on } \sum_D, (V^\perp.\nu)d_L \text{ on } \sum_L, d_{out} \text{ on } \sum_{out}\}$ implies that the source position S is the unique point of Ω subject to the nonlinear system :*

$$\begin{cases} \psi(S) = \ln\left(\dfrac{P_\psi}{P_0}\right) \\ \psi^\perp(S) = \dfrac{P_{\psi^\perp}}{P_0} \end{cases} \text{ and } \lambda(.) \text{ is subject to : } \int_0^{T^0} e^{Rt}\lambda(t)dt = e^{RT}P_0 \quad (IV.54)$$

Where ψ, $\psi^\perp$ are the two functions defined in (IV.18), (IV.21) and P_0, P_ψ, $P_{\psi^\perp}$ are such that

$$\begin{aligned}
P_0 &= \int_\Omega u(.,T)d\Omega + \int_{\Gamma_{out}\times(0,T)} e^{-R(T-t)}d_{out}V.\nu d\Gamma dt - \int_{\Gamma_D\times(0,T)} e^{-R(T-t)}d_{in}d\Gamma dt \\
P_\psi &= \int_\Omega e^\psi u(.,T)d\Omega - \int_{\Gamma_D\times(0,T)} e^{\psi-R(T-t)}d_{in}d\Gamma dt \\
P_{\psi^\perp} &= \int_\Omega \psi^\perp u(.,T)d\Omega + \int_{\Gamma_{out}\times(0,T)} e^{-R(T-t)}d_{out}\left(\psi^\perp V - V^\perp\right).\nu d\Gamma dt \\
&\quad - \int_{\Gamma_D\times(0,T)} e^{-R(T-t)}\psi^\perp d_{in}d\Gamma dt - \int_{\Gamma_L\times(0,T)} e^{-R(T-t)}(V^\perp.\nu)d_L d\Gamma dt
\end{aligned} \quad (IV.55)$$

Proof. Since (IV.12) holds true then, from multiplying (IV.1)-(IV.2) by z that solves the adjoint equation (IV.23) and integrating by parts over Ω using the initial-boundary conditions (IV.7) with integrating the obtained result over $(0,T)$ we find

$$\begin{aligned}
\int_0^{T^0} \lambda(t)z(S,t)dt &= \int_\Omega z(.,T)u(.,T)d\Omega + \int_{\Gamma_{out}\times(0,T)} u\left(zV + D\nabla z\right).\nu d\Gamma dt \\
&\quad - \int_{\Gamma_D\times(0,T)} zD\nabla u.\nu d\Gamma dt + \int_{\Gamma_L\times(0,T)} uD\nabla z.\nu d\Gamma dt
\end{aligned} \quad (IV.56)$$

Moreover, by substituting in (IV.56) z by the three functions $z_g(x,t) = e^{g(x)-R(T-t)}$ for $g = 0, \psi$ and $z^\perp(x,t) = \psi^\perp(x)e^{-R(T-t)}$ as in (IV.22) that solve the adjoint equation (IV.23) in $\Omega \times (0,T)$ and satisfy on $\Gamma_L \times (0,T) : D\nabla z_g.\nu = 0$ and $D\nabla z^\perp.\nu = -e^{-R(T-t)}V^\perp.\nu$, we get

Chapitre IV. Inverse source problem based on two dimensionless dispersion-current functions in 2D evolution transport equations with spatially varying coefficients

$$\overline{\lambda} = \int_\Omega u(.,T)d\Omega + \int_{\Gamma_{out}\times(0,T)} e^{-R(T-t)}uV.\nu d\Gamma dt - \int_{\Gamma_D\times(0,T)} e^{-R(T-t)}D\nabla u.\nu d\Gamma dt$$
$$\overline{\lambda}e^{\psi(S)} = \int_\Omega e^\psi u(.,T)d\Omega - \int_{\Gamma_D\times(0,T)} e^{\psi-R(T-t)}D\nabla u.\nu d\Gamma dt \qquad \text{(IV.57)}$$
$$\overline{\lambda}\psi^\perp(S) = \int_\Omega \psi^\perp u(.,T)d\Omega + \int_{\Gamma_{out}\times(0,T)} e^{-R(T-t)}u\left(\psi^\perp V - V^\perp\right).\nu d\Gamma dt$$
$$- \int_{\Gamma_D\times(0,T)} e^{-R(T-t)}\psi^\perp D\nabla u.\nu d\Gamma dt - \int_{\Gamma_L\times(0,T)} e^{-R(T-t)}(V^\perp.\nu)u d\Gamma dt$$

Where $\overline{\lambda} = \int_0^{T^0} e^{-R(T-t)}\lambda(t)dt$. Hence, setting the boundary observation operator $M[F]$ introduced in (IV.10) to the boundary records given in (IV.11) leads, in view of (IV.57), to the results announced in (IV.54)-(IV.55). Furthermore, the uniqueness of the solution S for the nonlinear system (IV.54) is an immediate consequence from (IV.50)-(IV.51) of Theorem 3.1
□

Remark 4.2
- *If the time-dependent source intensity function $\lambda(.)$ admits a constant sign a.e. on $(0,T^0)$, say for example $\lambda(t) \geq 0$, and the reaction coefficient $R \geq 0$ then, since we have $\lambda(t) \leq \lambda(t)e^{Rt} \leq \lambda(t)e^{RT^0}$ for all $t \in (0,T^0)$ and using (IV.54) we obtain*

$$e^{R(T-T^0)}P_0 \leq \int_0^{T^0} \lambda(t)dt \leq e^{RT}P_0 \qquad \text{(IV.58)}$$

In practice, (IV.58) could be of a great interest especially when the aim is rather to give an approximation/framing of the total amount loaded by the source function $\lambda(.)$
- *As the state u is subject to only knowledge of its boundary records, the localization of the source position S as well as the determination of lower and upper frame bounds for $\int_0^{T^0} \lambda(t)dt$ using (IV.54)-(IV.55) and (IV.58) are not so far possible since some terms in the coefficients P_0, P_ψ and $P_{\psi^\perp}$ are still involving the unknown data $u(.,T)$.*

In order to determine the three integrals in (IV.55) involving the unknown data $u(.,T)$, we establish the following result :

Proposition 4.3 *Let u be the solution of the problem (IV.1)-(IV.8), $T^0 \in (0,T)$ be such that (IV.12) holds true and $\gamma \in L^2\left(\Gamma_{out} \times (T^0,T)\right)$ be a boundary control that solves the boundary null controllability problem (IV.29)-(IV.30) for a given $\varphi_0 \in L^2(\Omega)$. Then, we have*

IV.4 Identification

$$\int_\Omega \varphi_0 u(.,T) d\Omega = \int_{\Gamma_D \times (T^0,T)} D\nabla u.\nu\varphi(.,T^0+T-t) d\Gamma dt \\ - \int_{\Gamma_{out} \times (T^0,T)} u\Big(D\nabla\varphi(.,T^0+T-t) + \gamma(.,T^0+T-t)V\Big).\nu d\Gamma dt \quad \text{(IV.59)}$$

Proof. Let φ be the solution of the problem (IV.29)-(IV.30). We use the change of variables : $\tilde{\varphi}(.,t) = \varphi(.,T^0+T-t)$ for all $t \in (T^0,T)$. Besides, since (IV.12) holds true, the solution u of (IV.1)-(IV.8) satisfies in $\Omega \times (T^0,T)$ a system similar to (IV.46) where the first equation becomes homogeneous and the initial state is $u(.,T^0)$. Thus, multiplying the first equation of this system by $\tilde{\varphi}$ and integrating by parts over Ω then, integrating the obtained result over (T^0,T) with using $\tilde{\varphi}(.,T^0) = \varphi(.,T) = 0$ and $\tilde{\varphi}(.,T) = \varphi(.,T^0) = \varphi_0$ in Ω gives

$$\int_\Omega \varphi_0 u(.,T) d\Omega = \int_{\Gamma_D \times (T^0,T)} \tilde{\varphi} D\nabla u.\nu d\Gamma dt - \int_{\Gamma_{out} \times (T^0,T)} u\Big(D\nabla\tilde{\varphi} + \tilde{\varphi}V\Big).\nu d\Gamma dt \quad \text{(IV.60)}$$

As we have $\tilde{\varphi}(.,t) = \gamma(.,T^0+T-t)$ on $\Gamma_{out} \times (T^0,T)$ then, the equation (IV.60) leads to the result announced in (IV.59). □

Applying Proposition 4.3 for the three initial states : $\varphi_0 = 1, e^\psi, \psi^\perp$ and using the boundary records (IV.10)-(IV.11) determines the terms in (IV.55) involving the unknown data $u(.,T)$.

4.1.1 Source localization procedure

According to (IV.54)-(IV.55) and assuming the coefficients $P_0, P_\psi, P_{\psi^\perp}$ obtained in (IV.55) to be now fully known, we localize the sought source position S as the unique solution of the following nonlinear system :

$$\psi(x) = \ln(\frac{P_\psi}{P_0}) \\ \psi^\perp(x) = \frac{P_{\psi^\perp}}{P_0} \quad \text{(IV.61)}$$

Where ψ and $\psi^\perp$ are the two dispersion-current functions obtained in (IV.18) and (IV.21). We remind that the uniqueness of the solution S for the nonlinear system (IV.61) is an immediate consequence of (IV.50)-(IV.51) from Theorem 3.1 Then, to determine the position S

Chapitre IV. Inverse source problem based on two dimensionless dispersion-current functions in 2D evolution transport equations with spatially varying coefficients

using Newton's method, for example, we compute the determinant $\det(Ja)$ of the 2×2 Jacobian matrix associated to the system (IV.61). In view of (IV.17) and (IV.19), that gives

$$\det\left(Ja(x)\right) = \partial_{x_1}\psi(x)\partial_{x_2}\psi^\perp(x) - \partial_{x_2}\psi(x)\partial_{x_1}\psi^\perp(x) = \frac{\|V(x)\|_2^2}{\det(D(x))}, \qquad \forall x \in \Omega \quad \text{(IV.62)}$$

Where D is the uniformly elliptic matrix introduced in (IV.4)-(IV.6) and $\|V\|_2^2 = V_1^2 + V_2^2 > 0$ from (IV.3). Therefore, the 2×2 Jacobian matrix associated to the nonlinear system (IV.61) is invertible in Ω. Thus, using (IV.17) and (IV.19) one can employ the following Newton's iterations to solve the system (IV.61) : Given an initial guess x^0 in Ω, compute

$$x^{k+1} = x^k + \frac{\det(D(x^k))}{\|V(x^k)\|_2^2}\begin{bmatrix} \frac{V_1}{D_M + D_T}(x^k)\left(\psi(x^k) - \ln(\frac{P_\psi}{P_0})\right) - \frac{V_2}{D_M + D_L}(x^k)\left(\psi^\perp(x^k) - \frac{P_{\psi^\perp}}{P_0}\right) \\ \frac{V_2}{D_M + D_T}(x^k)\left(\psi(x^k) - \ln(\frac{P_\psi}{P_0})\right) + \frac{V_1}{D_M + D_L}(x^k)\left(\psi^\perp(x^k) - \frac{P_{\psi^\perp}}{P_0}\right) \end{bmatrix} \quad \text{(IV.63)}$$

4.1.2 Computation of the HUM boundary control

As far as the boundary control γ required in Proposition 4.3 is concerned, we use the HUM boundary control obtained in (IV.35) of Theorem 2.1 To this end, we need to compute the minimizer $\hat{\xi}_0$ of the functional J introduced in (IV.34). Let G and G^* be the two following linear operators $G : L^2(\Omega) \longrightarrow L^2\left(\Gamma_{out} \times (T^0, T)\right)$ that to a given ξ_0 associates $G(\xi_0) = -D\nabla \xi.\nu$ where ξ is the solution of (IV.33) with the initial data $\xi(.,T) = \xi_0$, and $G^* : L^2\left(\Gamma_{out} \times (T^0, T)\right) \longrightarrow L^2(\Omega)$ that to a given f associates $G^*(f) = \Phi(.,T)$ where Φ is the solution of (IV.31) with $\varphi_0 = 0$ in Ω and $\Phi = f$ on $\Gamma_{out} \times (T^0, T)$. Then, according to (IV.37) we have for all $f \in L^2\left(\Gamma_{out} \times (T^0, T)\right)$ and all $\xi_0 \in L^2(\Omega)$

$$< G^*(f), \xi_0 >_{L^2(\Omega)} = < f, G(\xi_0) >_{L^2(\Gamma_{out} \times (T^0, T))} \quad \text{(IV.64)}$$

Hence, G and G^* are two adjoint operators. We introduce also the two linear operators $K : L^2(\Omega) \longrightarrow L^2(\Omega)$ that to a given ξ_0 associates $K(\xi_0) = \xi(.,T^0)$ where ξ is the solution of (IV.33) with the initial data $\xi(.,T) = \xi_0$, and $K^* : L^2(\Omega) \longrightarrow L^2(\Omega)$ that to a given Φ_0 associates $K^*(\Phi_0) = \Phi(.,T)$ where Φ is the solution of (IV.31) with $\gamma = 0$ on $\Gamma_{out} \times (T^0, T)$ and $\Phi(.,T^0) = \Phi_0$ in Ω. Then, using (IV.37) we obtain for all $\Phi_0 \in L^2(\Omega)$ and all $\xi_0 \in L^2(\Omega)$

IV.4 Identification

$$<K^*(\Phi_0), \xi_0>_{L^2(\Omega)} = <\Phi_0, K(\xi_0)>_{L^2(\Omega)} \tag{IV.65}$$

Which implies that K and K^* are also two adjoint operators. Therefore, in view of (IV.38) and using (IV.64)-(IV.65), the minimizer $\hat{\xi}_0$ of the functional J satisfies for all $\xi_0 \in L^2(\Omega)$

$$\begin{aligned}\nabla J(\hat{\xi}_0).\xi_0 &= <G(\hat{\xi}_0), G(\xi_0)>_{L^2(\Gamma_{out} \times (T^0, T))} - <e^{-\frac{1}{2}\psi}\varphi_0, K(\xi_0)>_{L^2(\Omega)} \\ &= <G^*G(\hat{\xi}_0) - K^*(e^{-\frac{1}{2}\psi}\varphi_0), \xi_0>_{L^2(\Omega)} \\ &= 0\end{aligned} \tag{IV.66}$$

Hence, we can determine the sought minimizer $\hat{\xi}_0$ of the functional J from solving :

$$G^*G(\hat{\xi}_0) = \Phi_{\varphi_0}(., T) \qquad \text{where} \quad G^*G : L^2(\Omega) \longrightarrow L^2(\Omega) \tag{IV.67}$$

And $\Phi_{\varphi_0}(., T) := K^*(e^{-\frac{1}{2}\psi}\varphi_0)$ that is the solution of the problem (IV.31) with $\gamma = 0$ on $\Gamma_{out} \times (T^0, T)$ taken at the final time $t = T$. To this end, we employ the so-called *Inner product method* [52] based on the use of a $N \times N$ matrix A as a representation of the linear controllability operator G^*G introduced in (IV.67). That transforms the computation of the minimizer $\hat{\xi}_0$ of the functional J into solving an associated linear system.

Proposition 4.4 *Let $(e_k)_{k \geq 1}$ be a complete orthonormal family of $L^2(\Omega)$. The $N \times N$ matrix A is symmetric positive semi-definite with entries*

$$A_{lk} = <D\nabla \xi_k.\nu, D\nabla \xi_l.\nu>_{L^2(\Gamma_{out} \times (T^0, T))} \qquad k, l = 1, .., N \tag{IV.68}$$

Where ξ_k and ξ_l are the solutions of (IV.33) with the initial data $\xi_k(., T) = e_k$ and $\xi_l(., T) = e_l$.

Proof. Let $X = \sum_{k \geq 1} x_k e_k$ and $Y = \sum_{k \geq 1} y_k e_k$. Provided X solves $AX = Y$, we get

$$\sum_{k \geq 1} x_k A e_k = \sum_{k \geq 1} y_k e_k \implies \sum_{k \geq 1} x_k <Ae_k, e_l>_{L^2(\Omega)} = y_l \qquad \text{for all} \quad l \geq 1$$

Chapitre IV. Inverse source problem based on two dimensionless dispersion-current functions in 2D evolution transport equations with spatially varying coefficients

Therefore, the entries of the matrix A are defined by $A_{lk} = <Ae_k, e_l>_{L^2(\Omega)}$, for all $k, l \geq 1$. Furthermore, since $e_k \in L^2(\Omega)$ for all $k \geq 1$ then, using the controllability operator introduced in (IV.67) and according to (IV.64), we obtain for all $k, l \geq 1$

$$A_{lk} = <Ae_k, e_l>_{L^2(\Omega)} = <G^*G(e_k), e_l>_{L^2(\Omega)} = <G(e_k), G(e_l)>_{L^2(\Gamma_{out} \times (T^0, T))} \qquad \text{(IV.69)}$$

Moreover, in view of (IV.69) and using the definition of the operator G, we find

$$Z^\top A Z = \left\| \sum_{k=1}^N z_k D \nabla \xi_k . \nu \right\|^2_{L^2(\Gamma_{out} \times (T^0, T))} \qquad \text{for all} \quad Z = (z_1, ..., z_N)^\top \in \mathbb{R}^N \qquad \text{(IV.70)}$$

Hence, from (IV.69) and (IV.70) we obtain the results announced in Proposition 4.4 □

Thus, in view of (IV.67) and using the inner product method, we determine the minimizer $\hat{\xi}_0$ of the functional J introduced in (IV.34) from solving the following associated linear system :

$$AX = Y \quad \text{with} \quad X = \sum_{k=1}^N <\hat{\xi}_0, e_k>_{L^2(\Omega)} e_k \quad \text{and} \quad Y = \sum_{k=1}^N <\Phi_{\varphi_0}(.,T), e_k>_{L^2(\Omega)} e_k \quad \text{(IV.71)}$$

Where A is the $N \times N$ symmetric positive semi-definite matrix defined in (IV.68) and $\Phi_{\varphi_0}(.,T)$ is the solution of the problem (IV.31) with $\gamma = 0$ on $\Gamma_{out} \times (T^0, T)$ taken at $t = T$.

Remark 4.5

– Due to the symmetry of the matrix A involved in the linear system (IV.71), only the computation of a half of its entries is needed.
– Solving the linear system obtained in (IV.71) leads to determine the HUM boundary control γ defined from (IV.35) that yields the boundary null controllability problem introduced in (IV.29)-(IV.30) with a given initial state $\varphi(.,T^0) = \varphi_0 \in L^2(\Omega)$. Therefore, resolving (IV.29)-(IV.30) for any new initial state $\varphi(.,T^0) = \hat{\varphi}_0 \in L^2(\Omega)$ requires only the update of the right-hand side Y in the linear system (IV.71).

Besides, as far as the complete orthonormal family $(e_k)_{k \geq 1}$ required in Proposition 4.4 is concerned, we introduce the following eigenvalue problem :

IV.4 Identification

$$\begin{aligned} -\operatorname{div}(D\nabla e) + \rho e &= \mu e & &\text{in } \Omega \\ \left(D\nabla e - \tfrac{1}{2} eV\right).\nu &= 0 & &\text{on } \Gamma_D \\ D\nabla e.\nu &= 0 & &\text{on } \Gamma_L \\ e &= 0 & &\text{on } \Gamma_{out} \end{aligned} \qquad (\text{IV.72})$$

Which leads to the associated spectral variational problem :

$$\begin{cases} \text{Find } \mu \in \mathbb{R} \text{ and } e \in \mathcal{V} \setminus \{0\} \text{ such that} \\ a(e,v) = \mu \int_\Omega evd\Omega \qquad \text{for all } v \in \mathcal{V} \end{cases} \qquad (\text{IV.73})$$

With $\mathcal{V} := \{v \in H^1(\Omega) \text{ such that } \left(D\nabla v - \tfrac{1}{2}vV\right).\nu\big|_{\Gamma_D} = 0, D\nabla v.\nu\big|_{\Gamma_L} = 0 \text{ and } v\big|_{\Gamma_{out}} = 0\}$ whereas $a(.,.)$ is the bilinear and symmetric form defined as follows :

$$a(e,v) = \int_\Omega D\nabla e \nabla v d\Omega + \int_\Omega \rho ev d\Omega - \frac{1}{2}\int_{\Gamma_D} evV.\nu d\Gamma \qquad (\text{IV.74})$$

Since $V.\nu \le 0$ on the inflow boundary Γ_D then, using the trace theorem and the uniform boundedness of the matrix D we prove the continuity of the bilinear form $a(.,.)$. In addition, the uniform ellipticity of the matrix D implies the coercivity of $a(.,.)$. Besides, according to the Embedding thoerem we have $\mathcal{V} \subset L^2(\Omega)$ with compact injection and $\mathcal{V}$ is dense in $L^2(\Omega)$. Therefore, the eigenvalues $(\mu_k)_{k\ge 1}$ of the variational spectral problem (IV.73) form an increasing sequence of positive real numbers that tends to infinity and the corresponding normalized eigenfunctions $(e_k)_{k\ge 1}$ which satisfy for all $k \ge 1$

$$e_k \in \mathcal{V}\setminus\{0\} \quad \text{and} \quad a(e_k, v) = \mu_k \int_\Omega e_k v d\Omega \qquad \text{for all } v \in \mathcal{V} \qquad (\text{IV.75})$$

Form a complete othonormal family of $L^2(\Omega)$.

Hence, in view of (IV.72) the solution ξ_k of the adjoint problem introduced in (IV.33) with the initial data $\xi_k(.,T) = e_k$ is given by $\xi_k(x,t) = e^{-\mu_k(T-t)}e_k(x)$. Then, according to Proposition 4.4, the entries A_{lk} of the matrix A can be computed as follows :

Chapitre IV. Inverse source problem based on two dimensionless dispersion-current functions in 2D evolution transport equations with spatially varying coefficients

$$A_{lk} = \int_{\Gamma_{out} \times (T^0, T)} e^{-(\mu_k + \mu_l)(T-t)} D\nabla e_k.\nu D\nabla e_l.\nu d\Gamma dt$$
$$= \frac{1 - e^{-(\mu_k + \mu_l)(T-T^0)}}{\mu_k + \mu_l} \int_{\Gamma_{out}} D\nabla e_k.\nu D\nabla e_l.\nu d\Gamma \tag{IV.76}$$

Furthermore, since the minimizer of the functional J is approximated by $\hat{\xi}_0$ determined from solving the linear system introduced in (IV.71) such that

$$\hat{\xi}_0(x) = \sum_{k=1}^{N} X_k e_k(x) \tag{IV.77}$$

Then, the solution $\hat{\xi}$ of the adjoint problem (IV.33) with the initial data $\hat{\xi}(.,T) = \hat{\xi}_0$ is given by $\hat{\xi}(x,t) = \sum_{k=1}^{N} e^{-\mu_k(T-t)} X_k e_k(x)$. Therefore, an approximations of the HUM boundary control defined from (IV.35) of Theorem 2.1 can be written as follows :

$$\gamma(x,t) = e^{\frac{1}{2}\psi(x)} \sum_{k=1}^{N} e^{-\mu_k(T-t)} X_k D\nabla e_k(x).\nu(x), \qquad \text{for } (x,t) \in \Gamma_{out} \times (T^0, T) \tag{IV.78}$$

Where $(X_1, .., X_N)^\top$ is the solution of the linear system introduced in (IV.71).

4.2 Identification of the time-dependent source intensity

In this subsection, we assume to be known the position S defining a time-dependent point source $F(x,t) = \lambda(t)\delta(x-S)$ that satisfies (IV.12) and focus on using the associated boundary observation operator $M[F]$ introduced in (IV.10) to identify the loaded time-dependent source intensity function $\lambda(.)$ To this end, we establish the following result :

Proposition 4.6 *Provided Lemma 4.1 applies for* $x^* = S$ *with* $\zeta_{in} \in L^2(\Gamma_D \times (0, T^0))$ *and* $\zeta_{out} \in L^2(\Gamma_{out} \times (0, T^0))$ *then,* $M[F] = \{d_{in} \text{ on } \Sigma_D, (V^\perp.\nu)d_L \text{ on } \Sigma_L, d_{out} \text{ on } \Sigma_{out}\}$ *implies that the sought function* $\lambda(.)$ *is subject to : for all* $\tau \in (0, T^0)$,

$$\int_0^\tau \lambda(t)v(S, \tau-t)dt = \int_{\Gamma_{out} \times (0,\tau)} d_{out}(x,t)\zeta_{out}(x, \tau-t)d\Gamma dt + \int_{\Gamma_L \times (0,\tau)} (V^\perp.\nu)d_L d\Gamma dt$$
$$- \int_{\Gamma_D \times (0,\tau)} d_{in}(x,t)\zeta_{in}(x, \tau-t)d\Gamma dt \tag{IV.79}$$

IV.4 Identification

Where v is the associated solution of the problem (IV.41).

Proof. Let v be the solution of the problem (IV.41) that satisfies (IV.42) with $x^* = S$ i.e., $v(S,.) \neq 0$ a.e. in $(0, T^0)$. Then, for all $\tau \in (0, T^0)$, multiplying the first equation in (IV.1)-(IV.8) by $\tilde{v}_\tau(.,t) = v(.,\tau - t)$ solution of (IV.52) and integrating by parts over Ω using Green's formula then, integrating the obtained result over $(0, \tau)$ gives

$$\int_0^\tau \lambda(t)\tilde{v}_\tau(S,t)dt = \int_{\Gamma_{out}\times(0,\tau)} u\left(D\nabla\tilde{v}_\tau + \tilde{v}_\tau V\right).\nu d\Gamma dt - \int_{\Gamma_D\times(0,\tau)} \tilde{v}_\tau D\nabla u.\nu d\Gamma dt \\ + \int_{\Gamma_L\times(0,\tau)} uD\nabla\tilde{v}_\tau.\nu d\Gamma dt \qquad \text{(IV.80)}$$

Besides, according to (IV.10), $M[F] = \{d_{in} \text{ on } \Sigma_D, (V^\perp.\nu)d_L \text{ on } \Sigma_L, d_{out} \text{ on } \Sigma_{out}\}$ implies

$$D\nabla u.\nu = d_{in} \text{ on } \Gamma_D \times (0,\tau), \ (V^\perp.\nu)u = (V^\perp.\nu)d_L \text{ on } \Gamma_L \times (0,\tau), \ u = d_{out} \text{ on } \Gamma_{out} \times (0,\tau) \text{(IV.81)}$$

As in view of (IV.52), we have $\tilde{v}_\tau = \zeta_{in}(.,\tau-t)$ on $\Gamma_D \times (0,\tau)$, $D\nabla\tilde{v}_\tau.\nu = V^\perp.\nu$ on $\Gamma_L \times (0,\tau)$ and $\left(D\nabla\tilde{v}_\tau(.,t) + \tilde{v}_\tau(.,t)V\right).\nu = \zeta_{out}(.,\tau-t)$ on $\Gamma_{out} \times (0,\tau)$ then, using (IV.81) in the right-hand side of the equation (IV.80) leads to the result announced in (IV.79). □

Hence, the identification of the unknown time-dependent intensity function $\lambda(.)$ can be transformed into solving the deconvolution problem (IV.79). To this end, given a desired number of time steps $\mathcal{M}$, we employ the regularly distributed discrete times $t_m = m\Delta t$, $m = 0,..,\mathcal{M}$ where $\Delta t = T/\mathcal{M}$. Then, assuming there exists $\mathcal{M}_0$ such that T^0 involved in (IV.12) satisfies $T^0 = \mathcal{M}_0 \Delta t$ and denoting $\lambda_m \approx \lambda(t_m)$, we find using the trapezoidal rule

$$\begin{aligned}\int_0^{t_{m+1}} \lambda(t)v(S,t_{m+1}-t)dt &= \sum_{k=0}^m \int_{t_k}^{t_{k+1}} \lambda(t)v(S,t_{m+1}-t)dt \\ &\approx \frac{\Delta t}{2}\sum_{k=0}^m \lambda_k v(S,t_{m+1-k}) + \lambda_{k+1}v(S,t_{m-k}) \qquad \text{(IV.82)} \\ &= \Delta t \sum_{k=1}^m \lambda_k v(S,t_{m+1-k})\end{aligned}$$

For all $m = 1,...,\mathcal{M}_0$, where according to (IV.41), we used $v(S,0) = 0$ and assumed that $\lambda_0 = 0$. Therefore, using (IV.82) we derive a discrete version of the deconvolution problem obtained in (IV.79) that leads to the following recursive formula :

Chapitre IV. Inverse source problem based on two dimensionless dispersion-current functions in 2D evolution transport equations with spatially varying coefficients

$$\lambda_m = \frac{1}{v(S,t_1)} \left(\frac{d_{m+1}}{\Delta t} - \sum_{k=1}^{m-1} \lambda_k v(S, t_{m+1-k}) \right), \quad \text{for all } m = 1, ..., \mathcal{M}_0 \quad \text{(IV.83)}$$

Where

$$d_m = \int_{\Gamma_{out} \times (0,t_m)} d_{out}(x,t) \zeta_{out}(x, t_m - t) d\Gamma dt + \int_{\Gamma_L \times (0,t_m)} (V^\perp . \nu) d_L(x,t) d\Gamma dt \\ - \int_{\Gamma_D \times (0,t_m)} d_{in}(x,t) \zeta_{in}(x, t_m - t) d\Gamma dt \quad \text{(IV.84)}$$

Then, for the clearness of our presentation we summarize the different steps of the established identification method in the following algorithm :

<u>Begin</u>

1. Compute $(e_k)_{k=1,..,N}$ and $(\mu_k)_{k=1,..,N}$ of the eigenvalue problem (IV.72).
   - Solve the problem (IV.31) with $\gamma = 0$ for $\varphi_0 = 1, e^\psi, \psi^\perp \implies \Phi_{\varphi_0=1,e^\psi,\psi^\perp}(0,T)$.
   - Solve the linear system $AX = Y$ in (IV.71) with $\Phi_{\varphi_0=1,e^\psi,\psi^\perp}(0,T)$.
   - Deduce from (IV.78) the HUM boundary control γ associated to $\varphi_0 = 1, e^\psi, \psi^\perp$.
   - Solve (IV.31) with γ associated to $\varphi_0 = 1, e^\psi, \psi^\perp \implies$ deduce (IV.59) for $\varphi_0 = 1, e^\psi, \psi^\perp$.

2. Compute the coefficients P_0, P_ψ and $P_{\psi^\perp}$ from (IV.55)-(IV.59).
   - Determine the sought source position S from (IV.63).
   - Determine the upper and lower frame bounds for $\int_0^T \lambda(t)dt$ from (IV.58).

3. Compute $\lambda_m \approx \lambda(t_m)$ from the recursive formula (IV.83).

<u>End</u>

5 Numerical experiments

We carry out some numerical experiments in the case of a rectangular domain defined by $\Omega := \{x = (x_1, x_2) \text{ such that } 0 < x_1 < L \text{ and } 0 < x_2 < \ell\}$ with the associated boundaries

IV.5 Numerical experiments

$$\begin{aligned}\Gamma_D &:= \{x = (x_1, x_2) \text{ such that } x_1 = 0 \text{ and } 0 < x_2 < \ell\} \\ \Gamma_{out} &:= \{x = (x_1, x_2) \text{ such that } x_1 = L \text{ and } 0 < x_2 < \ell\} \\ \Gamma_L &:= \{x = (x_1, x_2) \text{ such that } x_2 = 0 \text{ and } 0 < x_1 < L\} \\ &\cup \{x = (x_1, x_2) \text{ such that } x_2 = \ell \text{ and } 0 < x_1 < L\}\end{aligned} \qquad \text{(IV.85)}$$

In addition, we employ the velocity field $V = (V_1, V_2)^\top$ defined in Ω as follows :

$$V(x_1, x_2) = \begin{bmatrix} \overline{V}\left(\alpha e^{-\frac{\pi}{\ell}x_1} - \beta e^{\frac{\pi}{\ell}x_1}\right)\cos(\frac{\pi}{\ell}x_2) + V_0 \\ \overline{V}\left(\alpha e^{-\frac{\pi}{\ell}x_1} + \beta e^{\frac{\pi}{\ell}x_1}\right)\sin(\frac{\pi}{\ell}x_2) \end{bmatrix} \qquad \text{(IV.86)}$$

Where α, β and $\overline{V}, V_0$ are well selected real numbers. Then, V satisfies the required conditions introduced in (IV.3). Furthermore, using mean longitudinal and transverse dispersion coefficients D_L, D_T and in view of (IV.18), (IV.21) with $(a, b) = (0, 0)$, the two dispersion-current functions ψ and $\psi^\perp$ are rewritten as follows :

$$\begin{aligned}\psi(x_1, x_2) &= \frac{\ell}{\pi(D_M + D_L)}\left(\overline{V}\left(\alpha e^{-\frac{\pi}{\ell}x_1} + \beta e^{\frac{\pi}{\ell}x_1}\right)\cos(\frac{\pi}{\ell}x_2) - \frac{V_0\pi}{\ell}x_1 - \overline{V}(\alpha + \beta)\right) \\ \psi^\perp(x_1, x_2) &= -\frac{\ell}{\pi(D_M + D_T)}\left(\overline{V}\left(\alpha e^{-\frac{\pi}{\ell}x_1} - \beta e^{\frac{\pi}{\ell}x_1}\right)\sin(\frac{\pi}{\ell}x_2) + \frac{V_0\pi}{\ell}x_2\right)\end{aligned} \qquad \text{(IV.87)}$$

Besides, to derive the undimensionned version of the underlined inverse source problem, we introduce the variables : $x = x_1/L$, $y = x_2/\ell$ and $s = t/T$ that reduce the domain of study from $\Omega \times (0, T)$ into $(0, 1)^3$. As far as the discretization of the reduced domain is concerned, we employ uniform mesh sizes $\Delta x = 1/N_x$ and $\Delta y = 1/N_y$ with a constant time-step $\Delta s = 1/N_s$ and use a 5-points finite differences Crank-Nicholson scheme. To carry out numerical experiments, we use $L = 1000m$, $\ell = 100m$, $T = 14400s$ (4 hours) and $T^0 = 10800s$ (3 hours). In addition, we consider mean longitudinal and transverse dispersion coefficients $D_L = 10m^2s^{-1}$, $D_T = 0.2m^2s^{-1}$ with a molecular diffusion $D_M = 10^{-5}m^2s^{-1}$. Moreover, regarding the velocity field V and in view of (IV.86), we employ the coefficients $V_0 = 0.60ms^{-1}$, $\overline{V} = 10^{-2}$, $\alpha = 1.0$ and $\beta = e^{-\pi L/\ell}$. The reaction coefficient is $R = 10^{-5}s^{-1}$. We use a mesh with $N_x = 10$ and $N_y = 10$ whereas $N_s = 240$. To generate the boundary records d_{in}, d_L and d_{out} introduced in (IV.11), we solve using a 5-points finite differences Crank-Nicholson scheme the undimensionned version of the problem (IV.1)-(IV.7) with a source F as given in (IV.8) where the time-dependent

Chapitre IV. Inverse source problem based on two dimensionless dispersion-current functions in 2D evolution transport equations with spatially varying coefficients

intensity function is defined by

$$\lambda(t) = \sum_{n=1}^{3} c_n e^{-v_n(t-\tau_n)^2} \quad \text{if} \quad t \leq T^0 \quad \text{and} \quad 0 \text{ otherwise} \tag{IV.88}$$

Here, $c_1 = 1.2$, $c_2 = 0.4$, $c_3 = 0.6$ and $v_1 = 10^{-6}$, $v_2 = 5\ 10^{-5}$, $v_3 = 10^{-6}$. The coefficients τ_i are such that $\tau_1 = 4.5\ 10^3$, $\tau_2 = 6.5\ 10^3$, $\tau_3 = 9\ 10^3$. As far as the source position $S = (S_{x_1}, S_{x_2})$ is concerned, we employ the following approximation of the Dirac mass :

$$\delta(x_1 - S_{x_1}, x_2 - S_{x_2}) \approx \frac{1}{\pi \varepsilon^2} e^{-\frac{(x_1-S_{x_1})^2}{\varepsilon^2} - \frac{(x_2-S_{x_2})^2}{\varepsilon^2}} \tag{IV.89}$$

We set the parameter $\varepsilon = 10^{-5}$ in (IV.89). Then, to apply the identification method established in the previous section, we start by computing the three integrals in (IV.55) involving the unknown data $u(.,T)$. To this end and in view of Proposition 4.3, we solve using a 5-points finite differences Crank-Nicholson scheme the undimensionned version of the boundary null controllability problem introduced in (IV.29)-(IV.30) for the three following initial states : $\varphi_0 = 1, e^\psi, \psi^\perp$ in Ω by employing the HUM boundary control γ obtained in (IV.78) with $N = 10$. In the sequel, we present for each of those three cases the used initial state φ_0, the corresponding final state $\varphi(.,T)$ solution of the undimensionned version of the problem (IV.29) under the application of the associated HUM boundary control. For clearness, we present the obtained final state viewed from different positions on the reduced x-axis. We give also for each case the L^2-norm of the obtained final state.

IV.5 Numerical experiments

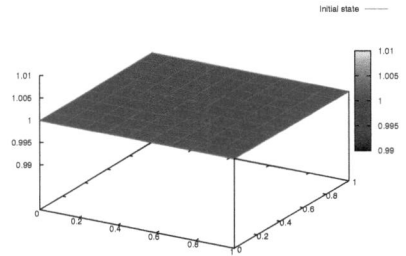

Figure 1: Initial state: $\varphi_0 = 1$

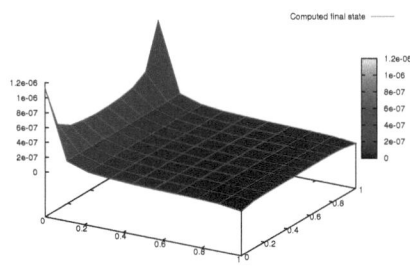

Final state: $\|\varphi(.,T)\|_{L^2((0,1)^2)} = 2.64E-07$

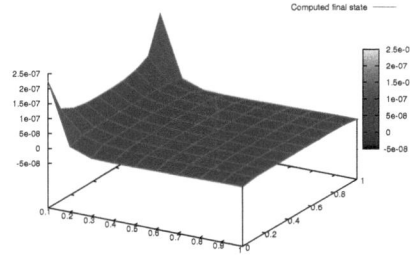

Figure 2: Final state for: $x \geq 0.1$

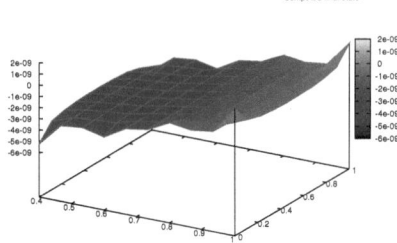

$x \geq 0.4$

Chapitre IV. Inverse source problem based on two dimensionless dispersion-current functions in 2D evolution transport equations with spatially varying coefficients

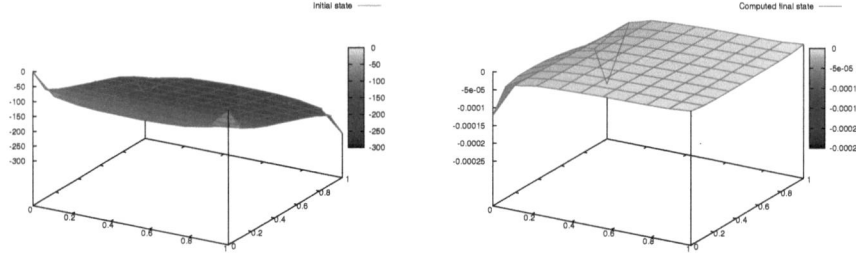

Figure 3: Initial state: $\varphi_0 = \psi^\perp$ Final state: $\|\varphi(.,T)\|_{L^2((0,1)^2)} = 4.06E{-}05$

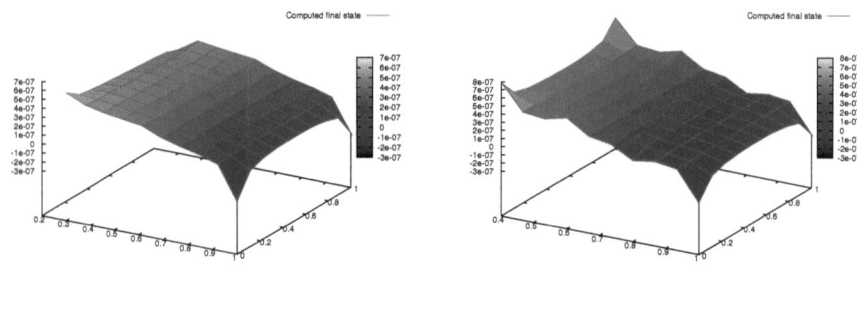

Figure 4: Final state for: $x \geq 0.3$ $x \geq 0.4$

IV.5 Numerical experiments

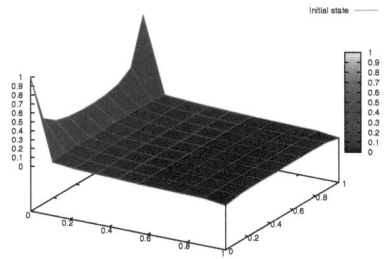
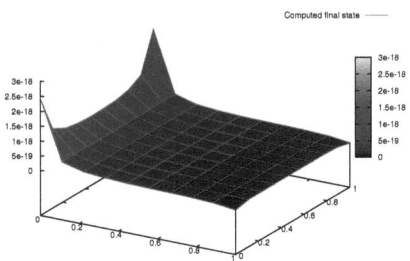

Figure 5: Initial state: $\varphi_0 = e^\psi$ Final state: $\|\varphi(.,T)\|_{L^2((0,1)^2)} = 5.76E{-}19$

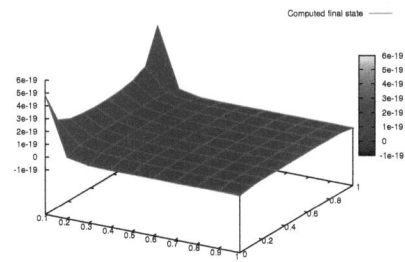
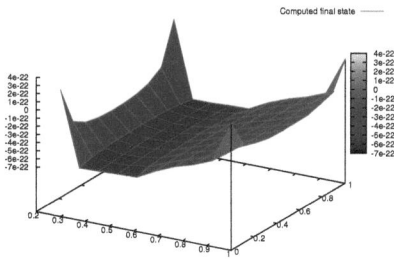

Figure 6: Final state for: $x \geq 0.1$ $x \geq 0.3$

Chapitre IV. Inverse source problem based on two dimensionless dispersion-current functions in 2D evolution transport equations with spatially varying coefficients

The analysis of the numerical results presented in Figures $1 - 6$ shows that the obtained HUM boundary control γ defined from (IV.35)-(IV.78) solves with relatively a good accuracy the boundary null controllability problem introduced in (IV.29)-(IV.30) for the three initial states $\varphi_0 = 1, e^\psi, \psi^\perp$. Then, using those results in Proposition 4.3 leads to determine the three integrals involving the unknown data $u(.,T)$ in (IV.55). Therefore, by employing the already generated boundary records d_{in}, d_L and d_{out} we compute the three coefficients P_0, P_ψ and $P_{\psi^\perp}$ obtained in (IV.55). Thus, as established in the previous section we localize the sought source position S as the unique solution of the nonlinear system introduced in (IV.54). To this end, we employ Newton's iterations defined in (IV.63). Here, we present some numerical results obtained from the localization of a source F as defined in (IV.8) emitting the same time-dependent intensity introduced in (IV.88) from different positions in Ω.

Used source position	Localized source position
S=(900,10)	S=(883.2,14.6)
S=(800,20)	S=(811.2,22.9)
S=(700,90)	S=(696.4,87.2)
S=(600,40)	S=(596.9,42.7)
S=(500,80)	S=(491.6,76.9)
S=(400,30)	S=(405.1,35.7)
S=(300,70)	S=(292.5,67.9)
S=(200,50)	S=(179.9,52.2)
S=(100,60)	S=(70.2,53.5)

Tableau IV.1 – Localization of a sought source position

The numerical results presented in Table 1 show that the source localization procedure established in the previous section enables to identify the sought source position with relatively a good accuracy. The observed error on the localized source position could be explained at least by the two following reasons : **1.** The used boundary records are not generated by a point source (Dirac mass) but rather by its Guassian approximation given in (IV.89) **2.** The sum of the error committed while solving the associated boundary null controllability problem (IV.29)-(IV.30) and the error on the approximation of integrals involved in the coefficients P_0, P_ψ and $P_{\psi^\perp}$ found in (IV.55).

In the remainder of this section, we aim to study numerically the stability of the identified results obtained from the established identification method. To this end, we introduce a Gaussian noise on the boundary records generated by a source located at $S = (600, 70)$ and loading the time-dependent intensity function defined in (IV.88). Then, for each introduced noise, we present on a same graphic the exact source intensity function given in (IV.88)

IV.5 Numerical experiments

And the identified one computed from the recursive formula obtained in (IV.83). In addition, we give also λ_{Error} that is the L^2 relative error on the identified source intensity function as well as the corresponding localized source position.

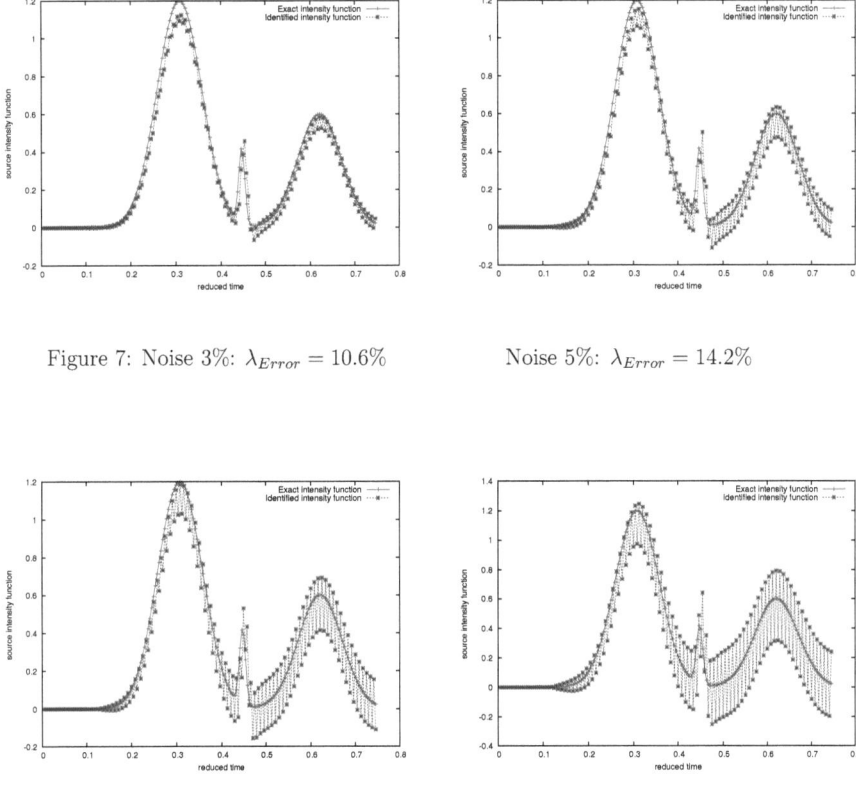

Figure 7: Noise 3%: $\lambda_{Error} = 10.6\%$ Noise 5%: $\lambda_{Error} = 14.2\%$

Figure 8: Noise 7%: $\lambda_{Error} = 21.7\%$ Noise 10%: $\lambda_{Error} = 36.1\%$

The corresponding localized source positions are the following: Noise 3%, $S = (591.3, 73.8)$; Noise 5%, $S = (582.6, 77.1)$; Noise 7%, $S = (617.2, 62.4)$; Noise 10%, $S = (553.5, 56.3)$.

Chapitre IV. Inverse source problem based on two dimensionless dispersion-current functions in 2D evolution transport equations with spatially varying coefficients

The identified results obtained from the established identification method for the elements S and $\lambda(.)$ defining the sought source F introduced in (IV.8) seem to be accurate and relatively stable with respect to the introduction of a Gaussian noise on the used boundary records. However, as far as the identified source intensity function $\lambda(.)$ is concerned, the analysis of the numerical results presented in Figures 7 − 8 seems indicating that the error tends to become relatively more significant as long as the time is going by. That could be explained by the fact that $\lambda(.)$ is identified from solving the deconvolution problem introduced in (IV.79). Therefore, for each $t_m \in (0, T^0)$ the identification of $\lambda(t_m)$ is affected by the noise introduced on the used boundary records only for the m first instances $k = 1, .., m$.

Conclusion

In this paper, we studied the nonlinear inverse source problem that consists of identifying from some boundary observations a time-dependent point source occurring in the right-hand side of an evolution $2D$ advection-dispersion-reaction equation with spatially varying coefficients. Under some reasonable assumptions, we established a constructive identifiability theorem proving that the elements defining the sought source are uniquely identified from some boundary observations related to the generated state. Those boundary observations are reduced to only recording the state on the outflow boundary and its flux on the inflow boundary of the monitored domain if the so-called no-slipping condition holds true. That led to establish a constructive identification method which localizes the sought source position as the unique solution of a nonlinear system of two equations, gives lower and upper frame bounds of the total amount loaded by the time-dependent source intensity function and transforms the identification of this latest into solving a deconvolution problem. Some numerical experiments on a variant of the surface water BOD pollution model were carried out. The obtained numerical results seem showing that the established identification method is accurate and relatively stable with respect to the introduction of a Gaussian noise on the used boundary records.

Chapitre V

Identification of the time active limit with lower and upper frame bounds of the total amount loaded by an unknown source in a 2D transport equation

Les résultats de ce chapitre sont issus de l'article [37], soumis pour publication en collaboration avec Adel Hamdi et Ahmed Rejaiba.

1 Introduction

In several areas of science and engineering, the mathematical modelling of a real-world problem leads to a partial/ordinary differential equation involving some active sources which could be unknown and even inaccessible. Therefore, getting more information about the involved active sources usually gives a better understanding of the ongoing physical phenomena and thus, could lead in some cases to take appropriate actions in order to prevent worse consequences. For example, that is the case for surface water pollution sources and their impact on the fauna, flora and human life. Inverse problems is a mathematical discipline that enables to learn more about those unknown sources provided some data on the consequences generated by their activities are available.

Although the state of the art concerning inverse source problems covers a broad spectrum of different source forms, it remains the main encountered difficulty while dealing with such kind of inverse problems is the no-identifiability (uniqueness) in general of a source in its abstract form, see [34] for a counterexample. In the literature, to overcome this difficulty authors generally assume available some *a priori* information on the form of the sought source : For example,

Chapitre V. Identification of the time active limit with lower and upper frame bounds of the total amount loaded by an unknown source in a 2D transport equation

time-independent sources are treated by Cannon J.R. in [12] using spectral theory, then by Engl H., Scherzer O. and Yamamoto M. in [21] using the approximated controllability of the heat equation. The results of this last paper are generalized by M. Yamamoto in [66] to sources of separated time and space variables where the time-dependent part is assumed to be known and null at the initial time then, recently in [52] to sources where the known time-dependent part of the sought source could also depend on the space variables and the involved differential operator is a time fractional parabolic equation. Hettlich F. and Rundell W. addressed in [38] the $2D$ inverse source problem in the heat equation where the sought source is the characteristic function associated to a subset of a disk. Hamdi A. and Mahfoudhi I. treated in [35, 36] the case of a time-dependent point source occurring in evolution transport equations where the source position and its time-dependent intensity function are both unknown.

The originality of this study consists in identifying, without any *a priori* information on the form of the involved unknown source, the time active limit that is the instant from which this unknown source occurring in a $2D$ evolution advection-dispersion-reaction equation becomes inactive. The identification method established in the present paper applied to some frequently encountered sources enables also to determine lower and upper frame bounds of the total amount loaded by the involved unknown sources without any *a priori* information neither on the number of active sources nor on their positions in the monitored domain nor on the form of their time-dependent intensity functions. A motivation to our study is a typical problem associated with environmental monitoring that consists of evaluating the surface water quality from recording the BOD (Biological Oxygen Demand) concentration which represents the amount of dissolved oxygen consumed by the micro-organisms during the oxidation process, see [4, 13, 23]. In a river, for example, identifying the time active limit together with lower and upper frame bounds of the total loaded pollution could lead to launch at time some first urgent actions : cleaning up the river, informing downstream drinking water stations, resuming some agriculture activities. In the literature, those informations are usually obtained from fully identifying the unknown involved pollution source which requires *a priori* information on its form to be available. In practice, that could delay and even limit the spectrum of launching such urgent actions. The paper is organized as follows : section 2 is devoted to stating the problem, assumptions and reminding some technical results for later use. In section 3, we prove the identifiability of the sought time active limit without any *a priori* information on the form of the involved unknown source from recording the generated state on the outflow boundary and its flux on the inflow boundary of the monitored domain. Section 4 is reserved to establish an identification method that uses those boundary records to determine the sought time active limit. Then, the

V.2 Mathematical modelling and prelimenary results

application of the established identification method to some frequently encountered sources leads to obtain also lower and upper frame bounds that provide a significant approximation of the total amount loaded by the involved unknown sources. Numerical experiments on a variant of the surface water BOD pollution model are presented in section 5.

2 Mathematical modelling and prelimenary results

Let $T > 0$ be a final monitoring time and Ω be a bounded and simply connected open subset of $\mathbb{R}^2$ with a sufficiently smooth boundary $\partial\Omega := \Gamma_D \cup \Gamma_N$. Here, Γ_D denotes the inflow boundary of the domain Ω whereas Γ_N regroups its two lateral boundaries Γ_L and its outflow boundary Γ_{out} i.e., $\Gamma_N := \Gamma_L \cup \Gamma_{out}$. The BOD concentration, denoted here by u, is governed by the following equation [13, 48, 49] :

$$L[u](x,t) = F(x,t) \quad \text{for all } (x,t) \in \Omega \times (0,T) \qquad (\text{V.1})$$

Where F represents the pollution source and L is the second-order linear parabolic partial differential operator defined as follows :

$$L[u](x,t) := \partial_t u(x,t) - \text{div}\left(D(x)\nabla u(x,t)\right) + V(x)\nabla u(x,t) + Ru(x,t) \qquad (\text{V.2})$$

With D is the hydrodynamic dispersion tensor, V is the flow velocity field and R is a real number that represents the first order decay reaction coefficient. Moreover, the velocity $V = \left(V_1, V_2\right)^\top$ is a spatially varying field that satisfies

$$\text{div}(V) = 0 \text{ in } \Omega \quad \text{and} \quad V.\nu = 0 \text{ on } \Gamma_L \qquad (\text{V.3})$$

Where ν is the unit outward normal vector to $\partial\Omega$. Hydrodynamic dispersion occurs as a consequence of two processes : molecular diffusion resulting from the random molecular motion and mechanical dispersion which is caused by non-uniform velocities. The summation of these two processes defines the hydrodynamic dispersion tensor, see [5] :

Chapitre V. Identification of the time active limit with lower and upper frame bounds of the total amount loaded by an unknown source in a 2D transport equation

$$D = D_M \mathbf{I} + \begin{bmatrix} D_1 & D_0 \\ D_0 & D_2 \end{bmatrix} \quad (V.4)$$

With $D_M > 0$ is a real number that represents the molecular diffusion coefficient, $\mathbf{I}$ is the 2×2 identity matrix and the spatially varying entries $D_{i=0,1,2}$ are such that, see [48, 49] :

$$D_1 = \frac{D_L V_1^2 + D_T V_2^2}{\|V\|_2^2}, \quad D_0 = \frac{V_1 V_2 (D_L - D_T)}{\|V\|_2^2} \quad \text{and} \quad D_2 = \frac{D_L V_2^2 + D_T V_1^2}{\|V\|_2^2} \quad (V.5)$$

Where $\|V\|_2 = \sqrt{V_1^2 + V_2^2}$ and D_T, D_L are the transverse and longitudinal dispersion coefficients that satisfy $0 \leq D_T < D_L$. Therefore, according to (V.4)-(V.5) the dispersion tensor D can be rewritten as follows :

$$D = (D_M + D_T)\mathbf{I} + \frac{D_L - D_T}{\|V\|_2^2} V V^\top \implies (D_M + D_T)\|X\|_2^2 \leq DX.X \leq (D_M + D_L)\|X\|_2^2 \quad (V.6)$$

For all $X \in \mathbb{R}^2$. Hence, (V.6) implies that the matrix D uniformly elliptic and bounded in Ω. In the remainder, we assume V_1, V_2 and D_L, D_T to be Lipschitz functions in Ω.

Furthermore, to (V.1)-(V.2) one has to append initial and boundary conditions. For the initial condition, we could use without loss of generality no pollution occurring at the initial monitoring time and thus, a null initial *BOD* concentration. As far as the boundary conditions are concerned, an homogeneous Dirichlet condition on the inflow boundary seems to be reasonable since the convective transport generally dominates the diffusion process. However, other physical considerations suggest to use rather a Neumann homogeneous condition on the remaining parts of the boundary. Then, we employ the following :

$$\begin{aligned} u(.,0) &= 0 & &\text{in } \Omega \\ u &= 0 & &\text{on } \Sigma_D := \Gamma_D \times (0,T) \\ D\nabla u.\nu &= 0 & &\text{on } \Sigma_N := \Gamma_N \times (0,T) \end{aligned} \quad (V.7)$$

Note that due to the linearity of the operator L introduced in (V.2) and according to the superposition principle, the use of a non-zero initial condition and/or inhomogeneous boundary

V.2 Mathematical modelling and prelimenary results

conditions do not affect the results established in this paper.

As far as the set of admissible sources is concerned, we call admissible all source F for which there exists $T^0 \in (0, T)$ such that

$$F(x,t) = f(x,t)\chi_{(0,T^0)}(t) \text{ and } \forall h \in \mathcal{E},\ \prec f(.,t), h(.,t) \succ > 0 \text{ a.e. in } (0, T^0) \quad (V.8)$$

With $\chi_{(0,T^0)}(t) = 1$ if $t \in (0, T^0)$ and 0 otherwise whereas here and in the remainder of this paper, $\prec .,. \succ$ represents the product in the distribution sense. Furthermore,

$$\mathcal{E} := \left\{ h : \Omega \times (0, T^0) \longrightarrow \mathbb{R} \text{ regular enough and } \forall x \in \Omega, h(x,.) > 0 \text{ a.e. in } (0, T^0) \right\} \quad (V.9)$$

For example, the following most encountered forms of surface water pollution sources :

$$F(x,t) = \sum_{k=1}^{K} \lambda_k(t)\delta(x - S^k) \quad \text{and} \quad F(x,t) = \sum_{k=1}^{K} \lambda_k(t)\chi_{\omega_k}(x) \quad (V.10)$$

Where δ denotes the Dirac mass, $K \geq 1$ and for $k = 1,..,K$ the elements defining F satisfy $\lambda_k(.) > 0$ a.e. in $(0, T^0)$, $\lambda_k(t) = 0$ for all $t \geq T^0$, $S^k \in \Omega$ and $\omega_k \subset \Omega$, are admissible.

Besides, given a source F regular enough and satisfying (V.8)-(V.9) with $T^0 \in (0, T)$, the problem (V.1)-(V.7) admits a unique solution u that is generally sufficiently smooth on $\partial\Omega$ which allows to define the following boundary observation operator :

$$M[T^0] := \left\{ D\nabla u.\nu \text{ on } \Sigma_D := \Gamma_D \times (0, T),\ u \text{ on } \Sigma_{out} := \Gamma_{out} \times (0, T) \right\} \quad (V.11)$$

This is the so-called *forward problem*. The *inverse problem* with which we are concerned here is : given some boundary records d_{in} of $D\nabla u.\nu$ on Σ_D and d_{out} of u on Σ_{out}, determine the sought time active limit $T^0 \in (0, T)$ subject to (V.8)-(V.9) that yields

$$M[T^0] = \left\{ d_{in} \text{ on } \Sigma_D,\ d_{out} \text{ on } \Sigma_{out} \right\} \quad (V.12)$$

For later use and as in view of (V.6) the matrix D is invertible in Ω, there exists a unique

Chapitre V. Identification of the time active limit with lower and upper frame bounds of the total amount loaded by an unknown source in a 2D transport equation

vector field X solution of the linear system $DX + V = 0$ in Ω. Moreover, we have

$$D^{-1} = \frac{1}{\det(D)}\begin{bmatrix} D_M + D_2 & -D_0 \\ -D_0 & D_M + D_1 \end{bmatrix} \implies X = -\frac{1}{\det(D)}\begin{bmatrix} D_M V_1 + D_2 V_1 - D_0 V_2 \\ D_M V_2 + D_1 V_2 - D_0 V_1 \end{bmatrix} \quad \text{(V.13)}$$

Furthermore, according to (V.5) we find

$$\begin{aligned} \det(D) &= (D_M + D_L)(D_M + D_T) \\ D_2 V_1 - D_0 V_2 &= V_1 D_T \\ D_1 V_2 - D_0 V_1 &= V_2 D_T \end{aligned} \quad \text{(V.14)}$$

Therefore, using the results (V.14) to substitute terms in the right-hand side of the second equation in (V.13) leads to $X = -\frac{1}{D_M + D_L} V$. Then, provided the following condition :

$$\text{rot}\left(\frac{V}{D_M + D_L}\right) = 0 \quad \text{(V.15)}$$

Holds true in Ω, the vector X is a gradient field derived from a scalar potential ψ which can be determined from solving

$$D\nabla\psi + V = 0 \quad \Leftrightarrow \quad \nabla\psi = -\frac{1}{D_M + D_L} V \quad \text{(V.16)}$$

Hence, ψ that satisfies $\psi(a, b) = 0$ where $(a, b) \in \Omega$, is defined as follows :

$$\psi(x_1, x_2) = -\int_a^{x_1} \left(\frac{V_1}{D_M + D_L}\right)(\eta, x_2) d\eta - \int_b^{x_2} \left(\frac{V_2}{D_M + D_L}\right)(a, \zeta) d\zeta \quad \text{(V.17)}$$

In addition, for later use we introduce the following adjoint system :

$$\begin{aligned} -\partial_t z - \text{div}(D\nabla z) - V\nabla z + Rz &= 0 \quad \text{in } \Omega \times (0, T) \\ z(., T) &\in L^2(\Omega) \\ D\nabla z.\nu &= 0 \quad \text{on } \Sigma_L := \Gamma_L \times (0, T) \end{aligned} \quad \text{(V.18)}$$

V.2 Mathematical modelling and prelimenary results

Provided (V.3) and (V.15) hold true, the two functions defined in $\Omega \times (0,T)$ as follows :

$$z_g(x,t) = e^{g(x)-R(T-t)} \quad \text{for } g = 0, \psi \tag{V.19}$$

Solve the system (V.18). Besides, we introduce the following boundary null-controllability problem that is to a given $\tau \in (0,T)$ and an initial state $\varphi_0 \in L^2(\Omega)$, determine a boundary control $\gamma \in L^2\big(\Gamma_{out} \times (\tau,T)\big)$ that drives the solution φ of the system :

$$\begin{aligned}
\partial_t \varphi - \text{div}(D\nabla\varphi) - V\nabla\varphi + R\varphi &= 0 & \text{in } \Omega \times (\tau,T) \\
\varphi(.,\tau) &= \varphi_0 & \text{in } \Omega \\
D\nabla\varphi.\nu &= 0 & \text{on } (\Gamma_D \cup \Gamma_L) \times (\tau,T) \\
\varphi &= \gamma & \text{on } \Gamma_{out} \times (\tau,T)
\end{aligned} \tag{V.20}$$

to satisfy : $\varphi(.,T) = 0$ in Ω \hfill (V.21)

To this end and provided the condition (V.15) holds true, let $\Phi = e^{-\frac{1}{2}\psi}\varphi$ in $\Omega \times (\tau,T)$ where ψ is the current-dispersion function defined in (V.17). Then, the boundary null-controllability problem introduced in (V.20)-(V.21) is equivalent to find a boundary control $\gamma \in L^2\big(\Gamma_{out} \times (\tau,T)\big)$ that leads the solution Φ of the following system :

$$\begin{aligned}
\partial_t \Phi - \text{div}(D\nabla\Phi) + \rho\Phi &= 0 & \text{in } \Omega \times (\tau,T) \\
\Phi(.,\tau) &= e^{-\frac{1}{2}\psi}\varphi_0 & \text{in } \Omega \\
\big(D\nabla\Phi - \tfrac{1}{2}\Phi V\big).\nu &= 0 & \text{on } \Gamma_D \times (\tau,T) \\
D\nabla\Phi.\nu &= 0 & \text{on } \Gamma_L \times (\tau,T) \\
\Phi &= e^{-\frac{1}{2}\psi}\gamma & \text{on } \Gamma_{out} \times (\tau,T)
\end{aligned} \tag{V.22}$$

to satisfy $\Phi(.,T) = 0$ in Ω \hfill (V.23)

Where $\rho = R + \frac{1}{4}V^\top D^{-1}V = R + \|V\|_2^2/4(D_M + D_L)$. Furthermore, we introduce the adjoint problem associated to (V.22) that is for a given $\xi_0 \in L^2(\Omega)$ determines ξ that solves

Chapitre V. Identification of the time active limit with lower and upper frame bounds of the total amount loaded by an unknown source in a 2D transport equation

$$\begin{cases} -\partial_t \xi - \text{div}(D\nabla \xi) + \rho \xi = 0 & \text{in } \Omega \times (\tau, T) \\ \xi(.,T) = \xi_0 & \text{in } \Omega \\ \left(D\nabla \xi - \frac{1}{2}\xi V\right).\nu = 0 & \text{on } \Gamma_D \times (\tau, T) \\ D\nabla \xi.\nu = 0 & \text{on } \Gamma_L \times (\tau, T) \\ \xi = 0 & \text{on } \Gamma_{out} \times (\tau, T) \end{cases} \quad (V.24)$$

And also the functional $J : L^2(\Omega) \to \mathbb{R}$ that is to a given ξ_0 associates

$$J(\xi_0) = \frac{1}{2} \prec D\nabla \xi.\nu, D\nabla \xi.\nu \succ_{L^2(\Gamma_{out} \times (\tau,T))} - \prec e^{-\frac{1}{2}\psi}\varphi_0, \xi(.,\tau) \succ_{L^2(\Omega)} \quad (V.25)$$

The authors in [36] determined for a problem similar to (V.22)-(V.23) the so-called HUM boundary control γ with refering to the Hilbert Uniqueness Method introduced by J. Lions [45, 46]. The result established in [36] is given by the following theorem :

Theorem 2.1 *Provided the spatially varying coefficients V_1, V_2 and D_L, D_T are Lipschitz functions that satisfy (V.15) then, for $\tau \in (0,T)$ and given $\varphi_0 \in L^2(\Omega)$ the boundary control*

$$e^{-\frac{1}{2}\psi}\gamma = D\nabla \hat{\xi}.\nu \quad \text{on} \quad \Gamma_{out} \times (\tau, T) \quad (V.26)$$

Is the control of smallest $L^2\big(\Gamma_{out} \times (\tau,T)\big)$-norm that leads the solution Φ of (V.22) to satisfy (V.23). Here, ψ is the dispersion-current function defined in (V.17) and $\hat{\xi}$ is the solution of (V.24) with $\hat{\xi}(.,T) = \hat{\xi}_0$ that is the unique minimizer of the functional J introduced in (V.25).

Moreover, using the *Inner product method* [52] with the normalized eigenfunctions $(e_k)_{k\geq 1}$ and the associated eigenvalues $(\mu_k)_{k\geq 1}$ solutions of the following eigenvalue problem :

$$\begin{cases} -\text{div}(D\nabla e) + \rho e = \mu e & \text{in } \Omega \\ \left(D\nabla e - \frac{1}{2}eV\right).\nu = 0 & \text{on } \Gamma_D \\ D\nabla e.\nu = 0 & \text{on } \Gamma_L \\ e = 0 & \text{on } \Gamma_{out} \end{cases} \quad (V.27)$$

The authors in [36] established an approximation of the HUM boundary control γ defined from (V.26) that leads, for a sufficiently large number K of initial terms, to employ

V.3 Identifiability

$$\gamma(x,t) = e^{\frac{1}{2}\psi(x)} \sum_{k=1}^{K} e^{-\mu_k(T-t)} X_k D\nabla e_k(x).\nu(x), \qquad \text{for } (x,t) \in \Gamma_{out} \times (\tau,T) \qquad (V.28)$$

Here, $X = (X_1,..,X_K)^\top$ is the solution of the following linear system : $AX = Y$ where the components of Y are $Y_k = \prec \Phi_{\varphi_0}(.,T), e_k \succ_{L^2(\Omega)}$ for $k = 1,..,K$ with $\Phi_{\varphi_0}(.,T)$ is the solution of the problem (V.22) for $\gamma = 0$ and taken at the final time $t = T$ whereas A is the $N \times N$ symmetric positive semi-definite matrix defined by

$$\begin{aligned} A_{lk} &= \int_{\Gamma_{out}\times(\tau,T)} e^{-(\mu_k+\mu_l)(T-t)} D\nabla e_k.\nu D\nabla e_l.\nu d\Gamma dt \\ &= \frac{1 - e^{-(\mu_k+\mu_l)(T-\tau)}}{\mu_k + \mu_l} \int_{\Gamma_{out}} D\nabla e_k.\nu D\nabla e_l.\nu d\Gamma \end{aligned} \qquad (V.29)$$

Remark 2.2
- *Due to the symmetry of the involved matrix A given in (V.29), only the computation of a half of its entries is needed.*
- *Solving the linear system $AX = Y$ leads to determine the HUM boundary control γ defined from (V.26)-(V.28) that yields the boundary null-controllability problem introduced in (V.20)-(V.21) for a given initial time $\tau \in (0,T)$. Therefore, resolving (V.20)-(V.21) for any new initial time $\tilde{\tau} \in (0,T)$ requires only the update of the entries A_{lk} i.e., substituting in (V.29) the coefficient $e^{-(\mu_k+\mu_l)(T-\tau)}$ by $e^{-(\mu_k+\mu_l)(T-\tilde{\tau})}$ and of the right hand-side Y i.e., substituting Y_k by $\tilde{Y}_k = \prec \tilde{\Phi}_{\varphi_0}(.,T), e_k \succ_{L^2(\Omega)}$ for $k = 1,..,K$ where $\tilde{\Phi}_{\varphi_0}(.,T)$ is the solution in $\Omega \times (\tilde{\tau},T)$ of (V.22) with $\gamma = 0$ and taken at $t = T$.*

3 Identifiability

In this section, provided the unknown source F involved in the problem (V.1)-(V.7) is admissible i.e., fulfills (V.8)-(V.9) with $T^0 \in (0,T)$ and thus, it becomes inactive for all $t \geq T^0$ we prove without any *a priori* assumption on the form of F that the boundary observation operator $M[T^0]$ introduced in (V.11) enables to determine uniquely the sought time active limit T^0. This result is given by the following theorem :

Theorem 3.1 *Let T_1^0, T_2^0 be two elements of $(0,T)$ and for $i = 1,2$, u_i be the solution of the problem (V.1)-(V.7) where the unknown source F satisfies (V.8)-(V.9) with $T^0 = T_i^0$. Then, provided the coefficients V_1, V_2 and D_L, D_T are Lipschitz functions in Ω, we have*

Chapitre V. Identification of the time active limit with lower and upper frame bounds of the total amount loaded by an unknown source in a 2D transport equation

$$M[T_1^0] = M[T_2^0] \implies T_1^0 = T_2^0 \qquad (V.30)$$

Where $M[T_i^0]$ is the boundary observation operator associated to u_i as introduced in (V.11).

Proof. Suppose that $T_1^0 \neq T_2^0$, for example, say $0 < T_1^0 < T_2^0 < T$ and let u_i be the solution of the problem (V.1)-(V.7) where F satisfies (V.8)-(V.9) with $T^0 = T_i^0$, for $i = 1, 2$. Then, the variable $w = u_2 - u_1$ solves

$$\begin{aligned}
L[w](x,t) &= f(x,t)\chi_{(T_1^0, T_2^0)}(t) & &\text{in } \Omega \times (0,T) \\
w(.,0) &= 0 & &\text{in } \Omega \\
w &= 0 & &\text{on } \Gamma_D \times (0,T) \\
D\nabla w . \nu &= 0 & &\text{on } \Gamma_N \times (0,T)
\end{aligned} \qquad (V.31)$$

Furthermore, in view of (V.11), having $M[T_1^0] = M[T_2^0]$ implies that

$$D\nabla w . \nu = 0 \quad \text{on } \Sigma_D \qquad \text{and} \qquad w = 0 \quad \text{on } \Sigma_{out} \qquad (V.32)$$

Then, multiplying the first equation in the problem (V.31) by a function z that satisfies the adjoint system introduced in (V.18) and integrating by parts over Ω with using the boundary conditions given in (V.31)-(V.32) leads to

$$\chi_{(T_1^0, T_2^0)}(t) \prec f(.,t), z(.,t) \succ = \frac{d}{dt} \prec w(.,t), z(.,t) \succ, \qquad \forall\, t \in (0,T) \qquad (V.33)$$

Besides, for $T^* \in (T_2^0, T)$ the variable w satisfies in $\Omega \times (T^*, T)$ the system (V.31) where the first equation becomes homogeneous and the initial condition is $w(., T^*)$. Therefore, according to (V.32) and using the unique continuation theorem from [42] it follows that we have $w(., T^*) = 0$ in Ω. Hence, by integrating the equation (V.33) over $(0, T^*)$ we obtain

$$\int_{T_1^0}^{T_2^0} \prec f(.,t), z(.,t) \succ dt = 0 \qquad (V.34)$$

Moreover, the function $z = z_g$ defined in (V.19) is a solution of the adjoint problem (V.18) and belongs to the set $\mathcal{E}$ introduced in (V.9). Thus, $\prec f(.,t), z_g(.,t) \succ > 0$ a.e. in (T_1^0, T_2^0) which implies that the equation (V.34) yields a contradiction. Therefore, it follows that $T_1^0 = T_2^0$.

$\square$

4   Identification

Given the boundary records introduced in (V.12) related to the state u solution of the problem (V.1)-(V.7) where the involved unknown source F satisfies (V.8)-(V.9) with $T^0 \in (0, T)$, we focus in this section on establishing an identification method that uses those records to determine without any *a priori* assumption on the form of F, the sought time active limit T^0. Furthermore, the application of the established identification method to the most encountered forms of surface water pollution sources gives also upper and lower frame bounds of the total amount loaded by the involved unknown sources. To this end, we start by proving the two following technical lemmas :

Lemma 4.1 *Let g be as introduced in (V.19) and F be a source regular enough that satisfies $\prec F, e^g \succ \in L^2(0, T)$. The solution u of the problem (V.1)-(V.7) with F is subject to*

$$
\begin{aligned}
\int_0^T \prec F(.,t), e^{g-R(T-t)} \succ dt &= \prec u(.,T), e^g \succ - \int_{\Gamma_D \times (0,T)} e^{g-R(T-t)} D\nabla u . \nu d\Gamma dt \\
&+ \int_{\Gamma_{out} \times (0,T)} e^{g-R(T-t)} u \big(D\nabla g + V \big) . \nu d\Gamma dt
\end{aligned}
\quad (V.35)
$$

Proof. From multiplying the equation (V.1)-(V.2) by z that solves the adjoint system introduced in (V.18) and integrating by parts over Ω using Green's formula and the boundary conditions satisfied by u we obtain for all $t \in (0, T)$,

$$
\prec F, z \succ = \frac{d}{dt} \prec u, z \succ + \int_{\Gamma_{out}} u \big(D\nabla z + zV \big) . \nu d\Gamma - \int_{\Gamma_D} z D\nabla u . \nu d\Gamma \quad (V.36)
$$

Therefore, from substituting in (V.36) z by z_g introduced in (V.19) that solves the adjoint system (V.18) and then, integrating the equation (V.36) over $(0, T)$, we find using the initial condition $u(.,0) = 0$ in Ω the result announced in (V.35). $\square$

Chapitre V. Identification of the time active limit with lower and upper frame bounds of the total amount loaded by an unknown source in a 2D transport equation

Besides, given g as introduced in (V.19) and $\tau \in (0,T)$, let $\gamma_g^\tau \in L^2\big(\Gamma_{out} \times (\tau,T)\big)$ be a boundary control that drives the solution φ_g^τ of the following system :

$$\begin{aligned}
\partial_t \varphi_g^\tau - \operatorname{div}(D\nabla \varphi_g^\tau) - V\nabla \varphi_g^\tau + R\varphi_g^\tau &= 0 && \text{in } \Omega \times (\tau,T) \\
\varphi_g^\tau(.,\tau) &= e^g && \text{in } \Omega \\
D\nabla \varphi_g^\tau . \nu &= 0 && \text{on } (\Gamma_D \cup \Gamma_L) \times (\tau,T) \\
\varphi_g^\tau &= \gamma_g^\tau && \text{on } \Gamma_{out} \times (\tau,T)
\end{aligned} \tag{V.37}$$

to satisfy : $\quad \varphi_g^\tau(.,T) = 0 \quad \text{in } \Omega$ \hfill (V.38)

Then, we establish the following technical lemma :

Lemma 4.2 *Let $\tau \in (0,T)$ and $\gamma_g^\tau \in L^2\big(\Gamma_{out} \times (\tau,T)\big)$ be a boundary control that solves the problem (V.37)-(V.38). Then, the solution u of the problem (V.1)-(V.7) yields*

$$\prec u(.,T), e^g \succ = \int_\tau^T \prec F(.,t), \varphi_g^\tau(.,\tau + T - t) \succ dt + \theta_g(\tau) \tag{V.39}$$

Where the mapping $\theta_g : (0,T) \longrightarrow \mathbb{R}$ that is to a given τ associates.

$$\begin{aligned}
\theta_g(\tau) &= -\int_{\Gamma_{out} \times (\tau,T)} u\Big(D\nabla \varphi_g^\tau(.,\tau+T-t) + \gamma_g^\tau(.,\tau+T-t)V\Big).\nu \, d\Gamma dt \\
&\quad + \int_{\Gamma_D \times (\tau,T)} D\nabla u.\nu \varphi_g^\tau(.,\tau+T-t) d\Gamma dt
\end{aligned} \tag{V.40}$$

Proof. Let $\gamma_g^\tau \in L^2\big(\Gamma_{out} \times (\tau,T)\big)$ be a solution of the boundary null-controllability problem introduced in (V.37)-(V.38). We employ on φ_g^τ the following change of variables : $\tilde{\varphi}_g^\tau(.,t) = \varphi_g^\tau(.,T+\tau-t)$ for all $t \in (\tau,T)$. Then, from multiplying the equation (V.1)-(V.2) by $\tilde{\varphi}_g^\tau$ and integrating by parts over Ω, we obtain for all $t \in (\tau,T)$,

$$\prec F, \tilde{\varphi}_g^\tau \succ = \frac{d}{dt} \prec u, \tilde{\varphi}_g^\tau \succ + \int_{\Gamma_{out}} u\Big(D\nabla \tilde{\varphi}_g^\tau + \tilde{\varphi}_g^\tau V\Big).\nu d\Gamma - \int_{\Gamma_D} \tilde{\varphi}_g^\tau D\nabla u.\nu d\Gamma \tag{V.41}$$

V.4 Identification

Therefore, by integrating both sides in the equation (V.41) over (τ, T) with using the final and initial states associated to $\tilde{\varphi}_g^\tau$ namely $\tilde{\varphi}_g^\tau(.,\tau) = 0$ and $\tilde{\varphi}_g^\tau(.,T) = e^g$ in Ω, we find

$$\begin{aligned} \prec u(.,T), e^g \succ &= \int_\tau^T \prec F(.,t), \tilde{\varphi}_g^\tau(.,t) \succ dt + \int_{\Gamma_D \times (\tau,T)} \tilde{\varphi}_g^\tau D\nabla u.\nu d\Gamma dt \\ &- \int_{\Gamma_{out} \times (\tau,T)} u\left(D\nabla \tilde{\varphi}_g^\tau + \tilde{\varphi}_g^\tau V\right).\nu d\Gamma dt \end{aligned} \quad (V.42)$$

Thus, from employing in the equation (V.42) the change of variables $\tilde{\varphi}_g^\tau(.,t) = \varphi_g^\tau(.,\tau+T-t)$ and the corresponding boundary condition $\tilde{\varphi}_g^\tau(.,t) = \gamma_g^\tau(.,\tau+T-t)$ on $\Gamma_{out} \times (\tau,T)$, we obtain the result announced in (V.39)-(V.40). □

In the sequel, using the two previous lemmas we transform the identification of the sought time active limit T^0 subject to (V.8)-(V.9) into an equivalent problem that consists of determining the instant from which a test function ζ_g defined from the boundary observations (V.11) becomes constant. This result is given by the following theorem :

Theorem 4.3 *Let g be as in (V.19), u be the solution of the problem (V.1)-(V.7) with a source F regular enough, θ_g be the mapping introduced in (V.40) and $\zeta_g : (0,T) \longrightarrow \mathbb{R}$ be such that to a given τ associates*

$$\begin{aligned} \zeta_g(\tau) &= \theta_g(\tau) + \int_{\Gamma_{out} \times (0,T)} e^{g-R(T-t)} u\left(D\nabla g + V\right).\nu d\Gamma dt \\ &- \int_{\Gamma_D \times (0,T)} e^{g-R(T-t)} D\nabla u.\nu d\Gamma dt \end{aligned} \quad (V.43)$$

Provided for all $\tau \in (0,T)$ there exists $\gamma_g^\tau \in L^2\left(\Gamma_{out} \times (\tau,T)\right)$ such that the solution φ_g^τ of the problem (V.37)-(V.38) satisfies a.e. in (τ,T), $\varphi_g^\tau(x,t) > \varphi_g^\tau(x,T) = 0$ for all $x \in \Omega$ then,

$$F \text{ fulfills } (V.8)\text{-}(V.9) \text{ with } T^0 \in (0,T) \quad \Leftrightarrow \quad \zeta_g'(\tau) = 0, \, \forall \tau \geq T^0 \quad (V.44)$$

Proof. Provided Lemma 4.2 applies then, from substituting $\prec u(.,T), e^g \succ$ in the equation (V.35) of Lemma 4.1 by its value given in (V.39)-(V.40), we obtain for all $\tau \in (0,T)$

$$\int_0^T \prec F(.,t), e^{g-R(T-t)} \succ dt - \int_\tau^T \prec F(.,t), \varphi_g^\tau(.,\tau+T-t) \succ dt = \zeta_g(\tau) \quad (V.45)$$

Chapitre V. Identification of the time active limit with lower and upper frame bounds of the total amount loaded by an unknown source in a 2D transport equation

Where ζ_g is the function introduced in (V.43). Therefore, if the source F satisfies (V.8)-(V.9) with $T^0 \in (0,T)$ which implies that $F(.,t) = 0$ in Ω for all $t \geq T^0$ then, the left-hand side in the equation (V.45) becomes a constant function with respect to τ for all $\tau \geq T^0$.

Now suppose that ζ_g introduced in (V.43) becomes a constant function for all $\tau \geq T^* \in (0,T)$ whereas the source F satisfies (V.8)-(V.9) with $T^0 \in (0,T)$ such that $T^0 \neq T^*$. Then, according to the first part of this proof, it follows that $T^0 > T^*$. Therefore, in view of (V.43) we get $\theta_g(\tau) = c$ for all $\tau \geq T^*$ where c is a real constant. Moreover, since the source F satisfies $F = 0$ in $\Omega \times (T^0, T)$ then, from applying Lemma 4.2 with $\tau \in (T^0, T)$ we obtain, as in (V.39), $\prec u(.,T), e^g \succ\, = c$. Hence, by reapplying Lemma 4.2 we obtain as found in (V.39) that

$$\int_\tau^{T^0} \prec F(.,t), \varphi_g^\tau(.,\tau+T-t) \succ dt = 0 \qquad \text{for all } \tau \in (T^*, T^0) \tag{V.46}$$

As by assumption, we have a.e. in (T^*, T), $\varphi_g^{T^*}(x,t) > \varphi_g^{T^*}(x,T) = 0$ for all $x \in \Omega$ and the source F satisfies (V.8)-(V.9) then, $\prec F(.,t), \varphi_g^{T^*}(.,t) \succ\, > 0$ a.e. in (T^*, T^0). Therefore, the equation (V.46) for $\tau = T^*$ yields a contradiction and thus, we have $T^* = T^0$. □

Thus, Theorem 4.3 transforms the task of identifying the sought time active limit $T^0 \in (0,T)$ associated to the unknown source F satisfying (V.8)-(V.9) that generated the boundary observations (V.11) into determining the instant T^0 from which the function ζ_g defined in (V.43) becomes constant in (T^0, T). Furthermore, as seen in the proof of Theorem 4.3, if ζ_g becomes a constant in (T^0, T) then, the function θ_g defined in (V.40) yields $\theta_g(\tau) = \prec u(.,T), e^g \succ$ for all $\tau \in (T^0, T)$. Therefore, the sought time active limit T^0 is redefined by the instant from which the function ζ_g yields

$$\begin{aligned}\zeta_g(\tau) &= \prec u(.,T), e^g \succ + \int_{\Gamma_{out} \times (0,T)} e^{g-R(T-t)} u\big(D\nabla g + V\big).\nu\, d\Gamma dt \\ &\quad - \int_{\Gamma_D \times (0,T)} e^{g-R(T-t)} D\nabla u.\nu\, d\Gamma dt, \qquad \text{for all } \tau \in (T^0, T)\end{aligned} \tag{V.47}$$

Remark 4.4 *Provided (V.15) holds true, the identification method applies with $g = 0$ and $g = \psi$. Although the option $g = 0$ seems simplifying computations, the use of $g = \psi$ could be more interesting especially in the case of high positive velocity components and/or small longitudinal dispersion coefficient (i.e., high Peclet number). In fact, according to (V.17) both situations lead to $|\psi|$ taking big values. Therefore, as with $a = b = 0$ in (V.17) we have $\psi < 0$ in Ω and the determination of the HUM boundary control obtained in (V.26)-(V.28) requires*

V.4 Identification

a symmetrization of the problem (V.20) which is done via the change of variables $\Phi = e^{-\frac{1}{2}\psi}\varphi$ that transforms the initial state from e^g into $e^{g-\frac{1}{2}\psi}$, the option $g = \psi$ could make the numerical resolution of the boundary null-controllability problem (V.37)-(V.38) much more easier since we would start from the initial state $0 < e^{\frac{1}{2}\psi} < 1$ rather than $e^{-\frac{1}{2}\psi} > 1$.

4.1 Application to some frequently encountered pollution sources

The application of the established identification method to the most encountered forms of surface water pollution sources introduced in (V.10) gives in addition to the identification of the sought time active limit T^0, lower and upper frame bounds of the total amount of pollution loaded by the involved unknown source F without any *a priori* information neither about the number N of active sources nor about their localizations and the forms of their time-dependent intensity functions. Thus, we obtain the following :

– Unknown time-dependent point sources i.e., $F(x,t) = \sum_{k=1}^{K} \lambda_k(t)\delta(x - S^k)$ where $K \geq 1$ and for $k = 1,..,K$, $\lambda_k \in L^2(0,T)$ with $\lambda_k(t) = 0$ for all $t \geq T^0 \in (0,T)$ and $\sum_{k=1}^{K} \lambda_k(t)$ admits a constant sign a.e. in $(0,T^0)$, say positive for example. Then, for a reaction coefficient $R \geq 0$ we get $\sum_{k=1}^{K} \lambda_k(t) \leq \sum_{k=1}^{K} \lambda_k(t)e^{Rt} \leq \sum_{k=1}^{K} \lambda_k(t)e^{RT^0}$ a.e. in $(0,T^0)$ which using the equation (V.45) with $g = 0$ leads to

$$\zeta_{g=0}(T^0)e^{R(T-T^0)} \leq \sum_{k=1}^{K} \int_{0}^{T^0} \lambda_k(t)dt \leq \zeta_{g=0}(T^0)e^{RT} \quad (V.48)$$

– Unknown time-dependent distributed sources i.e., $F(x,t) = \sum_{k=1}^{K} \lambda_k(t)\chi_{\omega_k}(x)$ where $K \geq 1$ and for $k = 1,..,K$, $\lambda_k \in L^2(0,T)$ with $\lambda_k(t) = 0$ for all $t \geq T^0 \in (0,T)$ and $\sum_{k=1}^{K} \lambda_k(t)\mathcal{A}_k$ admits a constant sign a.e. in $(0,T^0)$, say postive for example. Here, $\mathcal{A}_k$ designates the surface area of $\omega_k \subset \Omega$. Then, for a reaction coefficient $R \geq 0$ we have $\sum_{k=1}^{K} \lambda_k(t)\mathcal{A}_k \leq \sum_{k=1}^{K} \lambda_k(t)\mathcal{A}_k e^{Rt} \leq \sum_{k=1}^{K} \lambda_k(t)\mathcal{A}_k e^{RT^0}$ a.e. in $(0,T^0)$. Therefore, by employing $g = 0$ in the equation (V.45) we obtain

$$\zeta_{g=0}(T^0)e^{R(T-T^0)} \leq \sum_{k=1}^{K} \mathcal{A}_k \int_{0}^{T^0} \lambda_k(t)dt \leq \zeta_{g=0}(T^0)e^{RT} \quad (V.49)$$

Which given an estimation of the minimum and the maximum of $\mathcal{A}_k$ for $k = 1,..,K$ leads to lower and upper frame bounds of $\sum_{k=1}^{K} \int_{0}^{T^0} \lambda_k(t)dt$.

Chapitre V. Identification of the time active limit with lower and upper frame bounds of the total amount loaded by an unknown source in a 2D transport equation

5 Numerical experiments

In this section, we carry out some numerical experiments in the case of a rectangular domain $\Omega := \{x = (x_1, x_2) \text{ such that } 0 < x_1 < L \text{ and } 0 < x_2 < \ell\}$ with the boundaries

$$
\begin{aligned}
\Gamma_D &:= \{x = (x_1, x_2) \text{ such that } x_1 = 0 \text{ and } 0 < x_2 < \ell\} \\
\Gamma_{out} &:= \{x = (x_1, x_2) \text{ such that } x_1 = L \text{ and } 0 < x_2 < \ell\} \\
\Gamma_L &:= \{x = (x_1, x_2) \text{ such that } x_2 = 0 \text{ and } 0 < x_1 < L\} \\
&\cup \{x = (x_1, x_2) \text{ such that } x_2 = \ell \text{ and } 0 < x_1 < L\}
\end{aligned} \qquad (V.50)
$$

We use mean longitudinal and transverse coefficients D_L, D_T and a mean velocity vector $V = (V_1, V_2)^\top$ perpendicular to the inflow boundary Γ_D i.e., $V_2 = 0$ with $V_1 > 0$. Then, according to (V.5), the hydrodynamic dispersion tensor D introduced in (V.4)-(V.6) is reduced to the 2×2 diagonal matrix of entries $D_{11} = D_M + D_L$ and $D_{22} = D_M + D_T$. Besides, we consider the following time-dependent point sources :

$$
F(x,t) = \sum_{k=1}^{K} \lambda_k(t) \delta(x - S^k), \qquad \text{for all } x = (x_1, x_2) \in \Omega,\ t \in (0, T) \qquad (V.51)
$$

Where $K \geq 1$, δ denotes the Dirac mass and for $k = 1, .., K$, $S^k = (S^k_{x_1}, S^k_{x_2}) \in \Omega$ represents a source position whereas $\lambda_k \in L^2(0, T)$ designates its associated time-dependent intensity function. Furthermore, using the dimensional analysis method, see Fischer [24], the variable u_0 satisfying the following system :

$$
\begin{aligned}
\partial_t u_0 - \text{div}(D \nabla u_0) + V \nabla u_0 + R u_0 &= F(x,t) \quad \text{in } \mathbb{R}^2 \times (0,T) \\
u_0(.,0) &= 0 \quad \text{in } \mathbb{R}^2
\end{aligned} \qquad (V.52)
$$

With $V = (V_1, 0)^\top$ and $D = diag(D_{11}, D_{22})$ is defined by

$$
u_0(x,t) = \frac{1}{4\pi\sqrt{D_{11}D_{22}}} \sum_{k=1}^{K} \int_0^t \frac{\lambda_k(\eta)}{t-\eta} e^{-\frac{\left(x_1 - S^k_{x_1} - V_1(t-\eta)\right)^2}{4D_{11}(t-\eta)} - \frac{\left(x_2 - S^k_{x_2}\right)^2}{4D_{22}(t-\eta)} - R(t-\eta)} d\eta \qquad (V.53)
$$

Therefore, using $\hat{u}$ the solution of the following problem :

V.5 Numerical experiments

$$\begin{aligned}
\partial_t \hat{u} - \text{div}(D\nabla \hat{u}) + V\nabla \hat{u} + R\hat{u} &= 0 &&\text{in } \Omega \times (0,T) \\
\hat{u}(.,0) &= 0 &&\text{in } \Omega \\
\hat{u} &= -u_0 &&\text{on } \Sigma_D \\
D\nabla \hat{u}.\nu &= -D\nabla u_0.\nu &&\text{on } \Sigma_N
\end{aligned} \quad (\text{V}.54)$$

We determine the solution u of the main problem (V.1)-(V.7) with the mean velocity vector $V = (V_1, 0)^\top$, the associated hydrodynamic dispersion tensor $D = diag(D_{11}, D_{22})$ and the source F introduced in (V.51) as $u = \hat{u} + u_0$. Furthermore, the assertion (V.15) is satisfied and thus, the dispersion-current function ψ introduced in (V.16)-(V.17) is well defined. Then, we employ the HUM boundary control γ_g^τ obtained in (V.26)-(V.28). To this end, we need to determine the normalized eigenfunctions of the associated eigenvalue problem introduced in (V.27). Those functions are defined in the rectangular domain $\Omega = (0, L) \times (0, \ell)$ with $D = diag(D_{11}, D_{22})$ and $\rho = R + V_1^2/(4D_{11})$ as follows :

$$e_{mn}(x_1, x_2) = c_{mn} f_n(x_1) \cos\left(\frac{m\pi}{\ell} x_2\right) \qquad m, n \geq 0 \qquad (\text{V}.55)$$

Where $(f_n)_{n\geq 0}$ are the eigenfunctions of the following regular Sturm-Liouville problem :

$$\begin{aligned}
-D_{11} f_n''(x_1) &= \alpha_n f_n(x_1) &&\text{for } 0 < x_1 < L \\
f_n(L) &= -D_{11} f_n'(0) + \tfrac{V_1}{2} f_n(0) = 0
\end{aligned} \quad (\text{V}.56)$$

With $\alpha_n = \mu_{mn} - D_{22}(m\pi/\ell)^2 - \rho$. In addition, multiplying the first equation in (V.56) by f_n and integrating by parts over $(0, L)$ leads to

$$\alpha_n = \frac{D_{11} \|f_n'\|_{L^2(0,L)}^2 + \frac{V_1}{2} f_n^2(0)}{\|f_n\|_{L^2(0,L)}^2} > 0, \qquad \text{for all } n \geq 0 \qquad (\text{V}.57)$$

Therefore, according to (V.56)-(V.57) we obtain for all $n \geq 0$

$$f_n(x_1) = \sin\left(\frac{\sqrt{\alpha_n}}{\sqrt{D_{11}}}(L - x_1)\right) \quad \text{where } \alpha_n \text{ solves : } \tan\left(\frac{\sqrt{\alpha_n}}{\sqrt{D_{11}}} L\right) = -\frac{2D_{11}}{LV_1} \frac{\sqrt{\alpha_n}}{\sqrt{D_{11}}} L \quad (\text{V}.58)$$

Chapitre V. Identification of the time active limit with lower and upper frame bounds of the total amount loaded by an unknown source in a 2D transport equation

With $\alpha_0 > 0$. Then, once the coefficients α_n for $n \geq 0$ are determined from solving the second equation in (V.58), we deduce the eigenvalues μ_{mn} associated to the eigenfunctions e_{mn} such that $\mu_{mn} = \rho + \alpha_n + D_{22}(m\pi/\ell)^2$. For numerical purposes, let M and N be two sufficiently large integers. We employ the following notations :

$$e_k = e_{mn} \text{ and } \mu_k = \mu_{mn} \quad \text{where} \quad k = mN + n, \quad \text{for } n = 0,..,N-1; \quad m = 0,..,M-1$$
$$e_l = e_{pq} \text{ and } \mu_l = \mu_{pq} \quad \text{where} \quad l = pN + q, \quad \text{for } q = 0,..,N-1; \quad p = 0,..,M-1$$

Thus, given $k = 0,..,MN-1$ we identify its associated index m as the quotient resulting from the Euclidean division of k by N and its index n as the remainder. That leads to determine the kth eigenfunction $e_k = e_{mn}$ and the associated eigenvalue $\mu_k = \mu_{mn}$. Since the unit normal vector exterior to the outflow boundary Γ_{out} is $\nu = (1,0)^\top$ then, using the eigenfunctions $e_k = e_{mn}$ and $e_l = e_{pq}$ defined from (V.55)-(V.58), the entries A_{lk} introduced in (V.29) can be determined for all $l, k = 0,..,MN-1$ as follows :

$$\begin{aligned} A_{lk} &= D_{11}^2 \frac{1 - e^{-(\mu_l + \mu_k)(T-\tau)}}{\mu_l + \mu_k} \int_0^\ell \partial_{x_1} e_k(L, x_2) \partial_{x_1} e_l(L, x_2) dx_2 \\ &= D_{11}\sqrt{\alpha_n \alpha_q} \frac{1 - e^{-(\mu_l + \mu_k)(T-\tau)}}{\mu_l + \mu_k} c_{mn} c_{pq} \int_0^\ell \cos(\frac{m\pi}{\ell}x_2)\cos(\frac{p\pi}{\ell}x_2)dx_2 \\ &= \begin{cases} \ell D_{11} \dfrac{1 - e^{-(\mu_l + \mu_k)(T-\tau)}}{\mu_l + \mu_k}\sqrt{\alpha_n \alpha_q} c_{mn} c_{pq} & \text{if } m = p = 0 \\ \dfrac{\ell D_{11}}{2} \dfrac{1 - e^{-(\mu_l + \mu_k)(T-\tau)}}{\mu_l + \mu_k}\sqrt{\alpha_n \alpha_q} c_{mn} c_{pq} & \text{if } m = p \neq 0 \\ 0 & \text{if } m \neq p \end{cases} \end{aligned} \quad (V.59)$$

Hence, for each fixed row $l = pN + q$ of the matrix A and since $A_{lk} = 0$ if $m \neq p$, only the N elements A_{lk} with $k = pN + n$ for $n = 0,..,N-1$ are different to zero. In addition, as for a given p we have N successive rows that correspond to this same p namely $l = pN + q$ for $q = 0,..,N-1$ then, the matrix A is a block diagonal matrix where the M block matrices are for $m = 0,..,M-1$ the following $N \times N$ matrices : $A_{qn}^m = A_{mN+q,mN+n}$ for $q, n = 0,..,N-1$. Therefore, to solve the linear system $AX = Y$ introduced in (V.28)-(V.29) we compute for each $m = 0,..,M-1$ the N components $X^m = \left(X_{mN},...,X_{mN+N-1}\right)^\top$ of the solution $X = \left(X^0,...,X^m,...,X^{M-1}\right)^\top \in \mathbb{R}^{MN}$ from solving the linear system

$$A^m X^m = Y^m \quad \text{where} \quad Y_q^m = <\Phi_{e^g}(.,T), e_{mq}>_{L^2(\Omega)} \quad \text{for } q = 0,..,N-1 \quad (V.60)$$

V.5 Numerical experiments

Where, as in (V.28)-(V.29), $\Phi_{e^g}(.,T)$ is the solution of the problem (V.22) with $\varphi_0 = e^g$ and $\gamma = 0$, taken at the final time $t = T$. Besides, in view of (V.55) and using (V.58) we find

$$c_{mn} = \begin{cases} \dfrac{\sqrt{2}}{\sqrt{\ell}\beta_n} & \text{if } m = 0 \\ \dfrac{2}{\sqrt{\ell}\beta_n} & \text{if } m \neq 0 \end{cases} \qquad \text{where} \quad \beta_n = \sqrt{L - \dfrac{\sqrt{D_{11}}}{2\sqrt{\alpha_n}}\sin\left(2\dfrac{\sqrt{\alpha_n}}{\sqrt{D_{11}}}L\right)}, \quad n = 1,..,N$$

Which implies according to (V.59) that

$$A_{lk} = \begin{cases} 2D_{11}\dfrac{\sqrt{\alpha_n\alpha_q}}{\beta_n\beta_q}\dfrac{1 - e^{-(\mu_l+\mu_k)(T-\tau)}}{\mu_l + \mu_k} & \text{if } m = p \\ 0 & \text{if } m \neq p \end{cases}$$

And thus, for all $m = 0,..,M-1$ the $N \times N$ matrix A^m involved in the linear system introduced in (V.60) is defined as follows :

$$A_{qn}^m = 2D_{11}\dfrac{\sqrt{\alpha_n\alpha_q}}{\beta_n\beta_q}\left(\dfrac{1 - e^{-(\mu_{mq}+\mu_{mn})(T-\tau)}}{\mu_{mq} + \mu_{mn}}\right) \qquad \text{for } q,n = 0,..,N-1 \qquad (V.61)$$

Proposition 5.1 *For all $m = 0,..,M-1$, the $N \times N$ matrix A^m introduced in (V.61) is symmetric and positive definite.*

Proof. Since $\mu_{mq} = \rho + \alpha_q + D_{22}(m\pi/\ell)^2$ and $\mu_{mn} = \rho + \alpha_n + D_{22}(m\pi/\ell)^2$ then, in view of (V.61) the entries A_{qn}^m can be rewritten as follows :

$$A_{qn}^m = 2D_{11}\dfrac{\sqrt{\alpha_q\alpha_n}}{\beta_q\beta_n}\int_\tau^T e^{-2(\rho+D_{22}(m\pi/\ell)^2)(T-t)}e^{-(\alpha_q+\alpha_n)(T-t)}dt \qquad (V.62)$$

Furthermore, from (V.62) and for all vector $Z = \left(z_0,..,z_{N-1}\right)^\top \in \mathbb{R}^N$ we have

$$Z^\top A^m Z = 2D_{11}\int_\tau^T e^{-2(\rho+D_{22}(m\pi/\ell)^2)(T-t)}\left(\sum_{n=0}^{N-1}\dfrac{e^{-\alpha_n(T-t)}\sqrt{\alpha_n}}{\beta_n}z_n\right)^2 dt \geq 0 \qquad (V.63)$$

Moreover, as $D_{11} > 0$ then, it follows from (V.63) that $Z^\top A^m Z = 0$ is equivalent to

Chapitre V. Identification of the time active limit with lower and upper frame bounds of the total amount loaded by an unknown source in a 2D transport equation

$$\sum_{n=0}^{N-1} \frac{e^{-\alpha_n(T-t)}\sqrt{\alpha_n}}{\beta_n} z_n = 0 \quad \text{a.e. in } (\tau,T) \quad \Rightarrow \quad z_n = 0 \quad \text{for all } n = 0,..,N-1 \quad \text{(V.64)}$$

The implication in (V.64) is obtained from the fact that deriving $N-1$ times the first equation in (V.64) then, taking this equation as well as its derivatives at $t = T$, for example, leads to an homogeneous linear system where the unknown vector is made by the components $(\sqrt{\alpha_n}/\beta_n)z_n$ and the involved matrix is the transpose of the Vandermonde matrix of entries $V_{ij} = \alpha_i^j$ for $i,j = 0, N-1$. Hence, as $(\alpha_n)_n$ is a strictly increasing sequence of positive terms and $\beta_n \neq 0$ for all $n = 0,..,N-1$ we find the announced result. □

Besides, assuming now to be known the vector $X = \left(X^0,..,X^m,..,X^{M-1}\right)^\top \in \mathbb{R}^{MN}$ where for $m = 0,..,M-1$ the component $X^m = \left(X_{mN},..,X_{mN+N-1}\right)^\top \in \mathbb{R}^N$ is determined from solving the linear system $A^m X^m = Y^m$ introduced in (V.60), the HUM boundary control obtained from (V.28) that solves the boundary null-controllability problem introduced in (V.37)-(V.38) can be rewritten on $(0,\ell) \times (\tau,T)$ as follows :

$$\gamma_g^\tau(x_2,t) = -\sqrt{D_{11}}e^{-\frac{V_1 L}{2D_{11}}} \sum_{m=0}^{M-1} \left(\cos(\frac{m\pi}{\ell}x_2) \sum_{n=0}^{N-1} c_{mn} e^{-\mu_{mn}(T-t)}\sqrt{\alpha_n} X_{mN+n}\right)$$

Where $\mu_{mn} = \rho + \alpha_n + D_{22}(m\pi/\ell)^2$ with $\rho = R + V_1^2/(4D_{11})$ and α_n obtained from solving the second equation in (V.58).

To carry out some numerical experiments, we use $L = 1000m$, $\ell = 100m$, $V_1 = 0.01ms^{-1}$, $D_{11} = 30m^2s^{-1}$, $D_{22} = 0.01m^2s^{-1}$ and suppose monitoring the rectangular domain $\Omega = (0,L) \times (0,\ell)$ during $T = 14400s$ (4 hours). Furthermore, to generate the boundary observations introduced in (V.11) we solve (V.53)-(V.54) with the time-dependent point sources F defined in (V.51). As far as the time-dependent source intensity functions λ_k involved in (V.51) are concerned, we employ for the first set of experiments the following function [34] :

$$\lambda(t) = \begin{cases} \sum_{n=1}^{3} c_n e^{-a_n(t-b_n)^2} & \text{if } t < T^0 \in (0,T) \\ 0 & \text{otherwise} \end{cases} \quad \text{(V.65)}$$

Where $c_1 = 1.2$, $c_2 = 0.4$, $c_3 = 0.6$ and $a_1 = 10^{-6}$, $a_2 = 5\ 10^{-5}$, $a_3 = 10^{-6}$. The coefficients b_n are such that $b_1 = 4.5\ 10^3$, $b_2 = 6.5\ 10^3$, $b_3 = 9\ 10^3$. We derive the undimensioned version of

V.5 Numerical experiments

the underlined problem that reduces our study from the domain $\Omega \times (0,T)$ into the unit cube $(0,1)^3$. We use a discretization of the unit cube with $N_{x_1} = 10$ points following the reduced x_1-axis, $N_{x_2} = 5$ points following the reduced x_2-axis and $N_T = 240$ points following the reduced time axis. We employ a five-points finite differences method with the Crank-Nicolson scheme to solve the undimensioned version of the problems (V.37) and (V.54). Furthermore, we use $N = M = 3$ eigenfunctions and eigenvalues from (V.55). We carried out numerical tests with the option $g = 0$.

We start by presenting the first set of our numerical experiments that represents the state u in the reduced domain $(0,1)^2$ taken after an hour from the initial time and the curve of the associated function $\zeta_{g=0}$ introduced in (V.43) of Theorem 4.3 for each of the three following cases regarding the time-dependent point sources F introduced in (V.51) : Figure 1 corresponds to the case of an only one active point source i.e., $K = 1$ in (V.51), situated at $S = (600, 40)$ and emitting the source intensity function λ introduced in (V.65) with $T^0 = \frac{3}{4}T$ (3 hours). Figure 2 represents the case of two active point sources situated at $S^1 = (300, 60)$, $S^2 = (700, 40)$ and emitting the same source intensity function i.e., for $k = 1, 2$, $\lambda_k = \lambda$ introduced in (V.65) with $T^0 = \frac{1}{2}T$ (2 hours). Figure 3 describes the case of three active point sources situated at $S^1 = (300, 60)$, $S^2 = (500, 40)$, $S^3 = (700, 60)$ and emitting the same source intensity function i.e., for $k = 1, 2, 3$, $\lambda_k = \lambda$ introduced in (V.65) with $T^0 = \frac{3}{8}T$ (3/2 hour).

Chapitre V. Identification of the time active limit with lower and upper frame bounds of the total amount loaded by an unknown source in a 2D transport equation

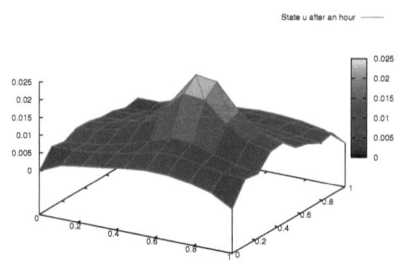
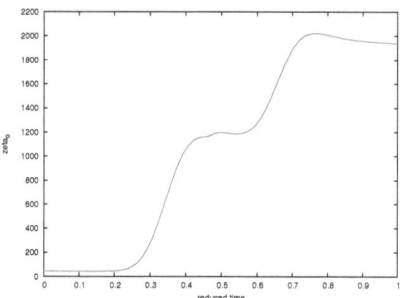

Figure 1: State u with $T^0 = \frac{3}{4}T$ Associated function $\zeta_{g=0}$

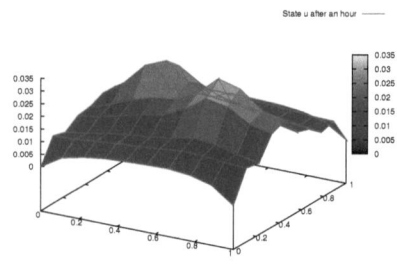
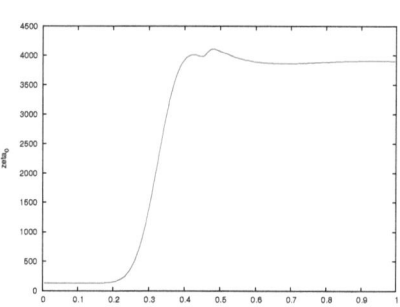

Figure 2: State u with $T^0 = \frac{1}{2}T$ Associated function $\zeta_{g=0}$

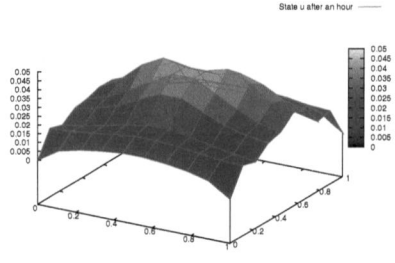
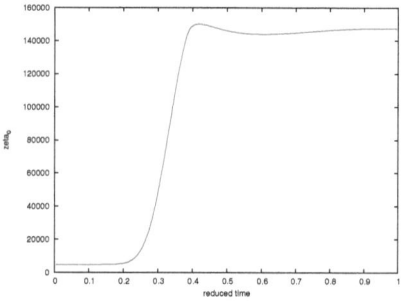

Figure 3: State u with $T^0 = \frac{3}{8}T$ Associated function $\zeta_{g=0}$

V.5 Numerical experiments

The analysis of the numerical results presented in Figures $1-3$ shows that the established identification method that determines the sought source time active limit T^0 as the instant from which the test function ζ_g defined in (V.43) becomes constant i.e., yields (V.47), enables to identify T^0 in the case of one or more active sources. However, it appears as a delay on the identified time active limit with respect to the used T^0 in (V.65) while generating the boundary observations (V.11). An explanation of this delay could be the fact that in general the convective transport dominates the diffusion process and thus, the significant information about the source activity is rather given by the data recorded on the outflow boundary. Therefore, the observed delay seems corresponding to the time needed by the last data emitted by the source F to reach the sensors on the outflow boundary.

In the remaining part of this paper, we aim to evaluate the lower and upper frame bounds obtained in (V.48) of the total amount loaded by the source F introduced in (V.51) for two different kinds of source intensity functions and various source positions. The results of this evaluation are summarized in the two following tables : numerical results presented in Table 1 correspond to the first used kind of source intensity functions that is for all $k = 1, .., K$, $\lambda_k(t) = 10^{-3}$ in $(0, T^0)$ and 0 otherwise, with the time active limit $T^0 = T/2$. That implies $\sum_{k=1}^{K} \int_0^{T^0} \lambda_k(t) dt = KT10^{-3}/2$. Besides, the numerical results given in Table 2 correspond to the second used kind of source intensity functions that is for all $k = 1, .., K$, $\lambda_k(t) = 10^{-3} \sin(\pi t/T^0)$ in $(0, T^0)$ and 0 otherwise, with the time active limit $T^0 = 3T/4$. Thus, we have $\sum_{k=1}^{K} \int_0^{T^0} \lambda_k(t) dt = 3KT10^{-3}/2\pi$.

Nb. source	Source position(s)	Frame bounds	$\sum_{k=1}^{K} \int_0^{T^0} \lambda_k(t) dt$
K=1	$S = (300, 60)$	[7.96, 8.55]	7.2
K=1	$S = (800, 60)$	[7.49, 8.02]	7.2
K=2	$S^1 = (300, 40)$ $S^2 = (800, 60)$	[14.65, 15.82]	14.4
K=2	$S^1 = (500, 40)$ $S^2 = (800, 60)$	[13.38, 14.37]	14.4
K=2	$S^1 = (700, 40)$ $S^2 = (800, 60)$	[13.54, 14.56]	14.4
K=3	$S^1 = (400, 60)$ $S^2 = (500, 80)$ $S^3 = (600, 20)$	[22.69, 24.39]	21.6
K=3	$S^1 = (500, 60)$ $S^2 = (600, 80)$ $S^3 = (700, 20)$	[21.97, 23.61]	21.6

Tableau V.1 – Frame bounds for the total amount loaded by the source F

Chapitre V. Identification of the time active limit with lower and upper frame bounds of the total amount loaded by an unknown source in a 2D transport equation

Nb. source	Source position(s)	Frame bounds	$\sum_{k=1}^{K} \int_{0}^{T^0} \lambda_k(t)dt$
K=1	$S = (300, 40)$	$[7.14, 7.95]$	6.87
K=1	$S = (900, 40)$	$[6.35, 7.08]$	6.87
K=2	$S^1 = (500, 60)$ $S^2 = (900, 20)$	$[13.79, 15.36]$	13.75
K=2	$S^1 = (700, 60)$ $S^2 = (900, 20)$	$[13.71, 15.28]$	13.75
K=3	$S^1 = (500, 80)$ $S^2 = (700, 40)$ $S^3 = (800, 20)$	$[20.62, 22, 98]$	20.62

Tableau V.2 – Frame bounds for the total amount loaded by the source F

The numerical experiments presented in Tables 1-2 show that the constant value $\zeta_{g=0}(T^0)$ reached by the function ζ_g introduced in (V.43) with $g = 0$ gives good lower and upper frame bounds, as established in (V.48), that yield a significant approximation of the total amount loaded by the time-dependent point sources F in the case of one or more active point sources and for the two used different kinds of source intensity functions. Nevertheless, from those experiments it appears that the accuracy on the established two frame bounds seems getting better as much as the active sources occur closer to the outflow boundary. In our opinion, this observation could be explained at least in part firstly, by the spread of pollution towards the two solid lateral boundaries with taking into account the slipping/no-slipping phenomena occurring on those boundaries and secondly, by the effect of chemical reactions along the way separating the source position from the sensors on the outflow boundary.

6 Conclusion

In this paper, we studied the identification of the time active limit associated to an unknown source occurring in the right-hand side of a $2D$ linear evolution advection-dispersion-reaction equation. We established an identification method using the records of the generated state on the outflow boundary and of its flux on the inflow boundary of the monitored domain that enables to determine the sought time active limit without any *a priori* information on the form of the involved unknown source. The application of the established identification method to the most encountered forms of surface water pollution sources gives in addition to the determination of the time active limit, lower and upper frame bounds of the total loaded pollution without any *a priori* information neither on the number of active sources nor on the form of their time-dependent intensity functions. We carried out some numerical experiments in the case of a rectangular domain with mean dispersion tensor and velocity field. The analysis of those experiments shows that the established identification method determines the sought time

V.6 Conclusion

active limit as well as the two frame bounds with a relatively good accuracy.

Discussion :

In practice, determining the time active limit together with lower and upper frame bounds of the total pollution amount loaded by the involved unknown source could be very interesting and even sufficient to engage first urgent actions, for example, in terms of surface water pollution : cleaning up the river, informing downstream drinking water stations, resuming some agriculture activities, etc .... Notice that usually the determination of such information is done by fully identifying the involved unknown source which is not always possible and requires *a priori* information on its form. That could delay and even limit the spectrum of launching such urgent actions.

The used HUM boundary control γ defined in (V.26) is obtained using the dispersion-current function ψ defined in (V.16)-(V.17) provided the assertion (V.15) holds true. However, if (V.15) is not fulfilled then, we denote ξ the solution of the adjoint problem associated to (V.20). Thus, ξ solves a problem similar to (V.24) where to the first equation $V\nabla\xi$ is added, and ρ is substituted by R whereas the boundary condition on $\Gamma_D \times (\tau, T)$ becomes $D\nabla\xi.\nu - \xi V.\nu = 0$. Therefore, provided V_1, V_2 and D_L, D_T are Lipschitz functions and using similar techniques as in [36], we prove that $\gamma = D\nabla\hat{\xi}.\nu$ is the HUM boundary control that solves (V.20)-(V.21). Here, likewise Theorem 2.1, $\hat{\xi}$ is the solution of the adjoint problem associated to (V.20) with the initial data $\hat{\xi}(.,T) = \hat{\xi}_0$ where $\hat{\xi}_0$ is the unique minimizer of the functional $J(\xi_0) = \frac{1}{2} \prec D\nabla\xi.\nu, D\nabla\xi.\nu \succ_{L^2(\Gamma_{out}\times(\tau,T))} - \prec \varphi_0, \xi(.,\tau) \succ_{L^2(\Omega)}$.

Conclusion générale et perspectives

Chapitre VI

Conclusion générale et perspectives

Dans ce livre, nous avons étudié les problèmes inverses de sources dans les équations aux dérivées partielles. Plus précisément, nous nous somme intéressés à l'identification des sources dépendantes du temps et de l'espace impliquées dans des équations de type advection-dispersion-réaction. Dans la pratique, ce type d'équations couvre un large spectre d'applications qui va de la biologie, à l'environnement avec l'étude de transport des polluants dans l'air, dans les eaux de surface et/ou souterraines. Le modèle considéré dans ce rapport porte sur la concentration de la demande biologique en oxygène (DBO) mais ne tient pas compte le déficit en oxygène [OD].

Premièrement, nous avons obtenu, des résultats d'identifiabilité, dans le cas monodimensionnel, pour le deux régimes stationnaire et transitoire ; par ailleurs nous avons introduit une méthode d'identification quasi-explicite permettant la localisation de la source recherchée et l'identification de sa fonction de débit à partir des mesures de l'état associé en deux points encadrant la zone source ; en outre la validation numérique montre que la méthode proposée offre une bonne précision et qu'elle est stable par rapport à l'introduction d'un bruit sur les mesures. Ces résultats ont fait l'objet d'un article déjà publié dans la revue internationale *Inverse Problems In Science Engineering*.

Nous avons étudié ensuite le cas bidimensionnel dans le même modèle de DBO, nous avons apporté des réponses positives aux questions d'identifiabilité de la source recherchée en mesurant le flux sur la frontière entrée et l'état sur la frontière sortie du domaine étudié. En outre, nous avons développé une méthode d'identification permettant de déterminer les éléments définissant la source qui est basée sur la technique de nulle-contrôlabilité frontière. J'ai validé numériquement cette méthode d'identification sur un modèle de la pollution des eaux de surface dans le cas bidimensionnel. Ces résultats ont fait l'objet d'un article soumis à une revue international.

Enfin, nous avons traité un nouveau type de problème inverse non-linéaire de source qui

Conclusion générale et perspectives

consiste en l'identification, via des mesures frontières, de l'instant limite à partir duquel une source dans sa forme abstraite a cessé ses émissions est devenue inactive. Les résultats de ce dernier chapitre ont fait l'objet d'un troisième article soumis dans la revue Inverse Problems and Imaging.

Prespectives

Les problèmes inverses font l'objet d'étude théorique mathématique. A ce jour, des nombreuses questions restent ouvertes, certaines d'entre elles étant étroitement liées aux plusieurs problèmes comme l'exemple étudié dans ce livre. Nous proposons dans ce qui suit quelques des pistes possibles pour les recherches.

1. Dans le cas mon-dimensionnel nous avons traité une équation linéaire à coefficients variables. un point qui reste à traiter, il s'agit du cas d'une équation quasi-linéaire et l'extension de ses résultats dans le modèle chaîné $DBO - OD$.

2. Dans le cas bidimensionnel nous avons (pu déterminer) des résultats pour une équation linéaire de type advection-diffusion-réaction avec des coefficients de régularité plus faible, ce qui reste dans ce cadre à traiter c'est l'identification d'une source ponctuelles au modèle couplé $DBO - OD$; et l'extension des résultats au cas d'une équation quasi-linéaire.

3. Quant aux travaux du dernier chapitre, ils nous ont inspiré, expressément, l'identification d'une temps limite d'une source devenue cesse d'émettre à partir d'un certain instant ; le dernière point qui reste à traiter le cas de l'interruption répétée des émissions. Mathématiquement cette source s'écrira sous la forme d'un somme des indicatrice en temps est leurs débits.

$$F(x,t) = \sum_{i=1}^{n} \chi_{(t_i,t_{i+1})}(t) f_i(x,t)$$

Aussi, la généralisation aux $3D$ pour les deux modèles simple et couplé, et ainsi élargissement du spectre d'application à d'autres problèmes physiques.

Chapitre VI. Conclusion générale et perspectives

Annexe A

Annexe

1 Chapitre 3

Proof of Proposition 5.1 For a constant velocity v_0 and a null reaction coefficient, the assertion (III.76) is equivalent to the main condition (III.15). And thus, according to Lemma 2.3, we have $\tilde{\varphi}_1(y) \neq 0$ for all $y \in (\tilde{a}, \tilde{b})$. Therefore, using (III.73) the function $\tilde{\Phi} : y \mapsto \tilde{\varphi}_2(y)/\tilde{\varphi}_1(y)$ is defined on $(\tilde{a}, \tilde{b})$ as follows :

$$\tilde{\Phi}(y) = \frac{\int_{\tilde{a}}^{y} \tilde{p}^{-1}(\eta)d\eta}{(c_1 - 1)\int_{\tilde{a}}^{y} \tilde{p}^{-1}(\eta)d\eta + c_1 \int_0^{\tilde{a}} \tilde{p}^{-1}(\eta)d\eta} \qquad (A.1)$$

$$= \frac{\Gamma(\tilde{b})\big(\Gamma(y) - \Gamma(\tilde{a})\big)}{\Gamma(\tilde{a})\big(\Gamma(\tilde{b}) - \Gamma(y)\big)}$$

Where $\Gamma(z) = \int_0^z \tilde{p}^{-1}(\eta)d\eta$. In addition, according to (III.74), we find

$$\Gamma(z) = d_m^{1-\frac{v_0}{\beta(d_m+\alpha)}} \int_0^z \left(d_m + \alpha(1 - e^{-\beta l\eta})\right)^{\frac{v_0}{\beta(d_m+\alpha)}-1} e^{\frac{lv_0}{d_m+\alpha}\eta} d\eta \qquad (A.2)$$

Then, using the change of variable $\xi = e^{-\beta l \eta}$ in (A.2) leads to

$$\Gamma(z) = \frac{d_m^{1-\frac{v_0}{\beta(d_m+\alpha)}}}{\beta l} \int_{e^{-\beta l z}}^{1} \left(\frac{d_m + \alpha(1-\xi)}{\xi}\right)^{\frac{v_0}{\beta(d_m+\alpha)}-1} \frac{d\xi}{\xi^2} \qquad (A.3)$$

Now, by employing the change of variable $\nu = 1/\xi$ in (A.3), we obtain

Annexe

$$\Gamma(z) = \frac{d_m}{lv_0}\left(\left(\frac{(d_m+\alpha)e^{\beta lz}-\alpha}{d_m}\right)^{\frac{v_0}{\beta(d_m+\alpha)}}-1\right) \quad (A.4)$$

Hence, using (A.4) in (A.1) gives the result announced in (III.77). □

Proof of Proposition 5.2 The function that given $y \in (0,1)$ associates $\int_0^y \tilde{D}^{-1/2}(z)dz$ is continuous and strictly increasing on $(0,\tilde{I})$. Then, using the change of variable $\xi = e^{-\beta lz}$ we obtain

$$\begin{aligned}
\tilde{\eta} &= \frac{1}{\beta\sqrt{T}}\int_{e^{-\beta ly}}^{1}\frac{1}{\xi\sqrt{d_m+\alpha(1-\xi)}}d\xi \\
&= \frac{1}{\beta\sqrt{T}(d_m+\alpha)}\int_{e^{-\beta ly}}^{1}\left(\frac{\sqrt{d_m+\alpha(1-\xi)}}{\xi}+\frac{\alpha}{\sqrt{d_m+\alpha(1-\xi)}}\right)d\xi \quad (A.5)\\
&= \frac{1}{\beta\sqrt{T(d_m+\alpha)}}\log\left(\frac{(d_m+\alpha)\left(1+\sqrt{1-\frac{\alpha}{d_m+\alpha}e^{-\beta ly}}\right)e^{\beta ly}-\frac{\alpha}{2}}{d_m\left(1+\sqrt{1+\frac{\alpha}{d_m}}\right)+\frac{\alpha}{2}}\right)
\end{aligned}$$

Therefore, from the last equality in (A.5), we find

$$\left(1+\sqrt{1-\frac{\alpha}{d_m+\alpha}e^{-\beta ly}}\right)e^{\beta ly} = \frac{\tau e^{\beta\sqrt{T(d_m+\alpha)}\tilde{\eta}}+\frac{\alpha}{2}}{d_m+\alpha} \quad \text{where } \tau = d_m\left(1+\sqrt{1+\frac{\alpha}{d_m}}\right)+\frac{\alpha}{2} \quad (A.6)$$

Multiplying and dividing the left side of the first equality in (A.6) by $1-\sqrt{1-(\alpha/(d_m+\alpha))e^{-\beta ly}}$ gives

$$\sqrt{1-\frac{\alpha}{d_m+\alpha}e^{-\beta ly}} = 1-\frac{\alpha}{\tau e^{\beta\sqrt{T(d_m+\alpha)}\tilde{\eta}}+\frac{\alpha}{2}}$$

Which leads to

$$y = \frac{2}{\beta l}\ln\left(\frac{\tau e^{\beta\sqrt{T(d_m+\alpha)}\tilde{\eta}}+\frac{\alpha}{2}}{\sqrt{2\tau(d_m+\alpha)}e^{\beta\sqrt{T(d_m+\alpha)}\tilde{\eta}/2}}\right) \quad (A.7)$$

Since from (A.6) we have $2\tau = (d_m + \alpha)(1 + \sqrt{d_m/(d_m + \alpha)})^2$, then using (A.6) in (A.7) we obtain the result announced in (III.79). □

2 Chapitre 4

Proof of Proposition 2.4

Here, we prove the result announced in Proposition 2.4 :

though
Références bibliographiques

Bibliographie

[1] Ainseba B. (1992) "Controlablité exacte, identifiabilité, sentinelles" Université de Technologie de Compiègne (thèse de doctorat). 17

[2] Andrle M.P, Ben Belgacem F., El Badia A. (2011) Identification of Moving Pointwise Sources in an Advection-Dispersion-Reaction Equation. Inverse Problems, 27 025007. 25

[3] Andrle M. and El Badia A. (2012) Identification of multiple moving pollution sources in surface waters or atmospheric media with boundary observations, Inverse Problems 28(2012) 075009. 25, 66

[4] APHA (1998) Standard Methods for the Examination of Water and Wastewater, 18th ed., American Public Health Association, Washington, DC. Bulut, V.N 23, 38, 100

[5] Bear J. and Bachmat Y. (1991) Introduction to Modeling of Transport Phenomena in Porous Media, Kluwer Academic Publishers, pp. 553 18, 67, 68, 101

[6] Baev A. (2005) Solution of the inverse dynamic problem of seismology with an unknown source, Computational Mathematics and Modeling, Vol. 2, Num. 3, p. 252-255 37

[7] Ben Belgacem F. (2012) Identifiability for the pointwise source detection in Fisher's reaction-diffusion equation, Inverse Problems, 28 065015

[8] Ben Belgacem F. (2012) Uniqueness for an ill-posed reaction-dispersion model. Application to organic pollution in stream-waters (One-dimensional model), Inverse Problems and Imaging 6-2, pp.163-181.

Références bibliographiques

[9] Brown L.c, T.O Barnwell "The Enhanced Stearm Water Qulity Models QUAL2E and QUAL2E-UNCAS : Documentation and user manual" (Environmental Reseach Loboratory Office of reseach and Developement U.S Environmental protection agency, Athens, Geogia 30613). 15, 16, 17, 20

[10] Bruce Stewart H. (1980) Generation of analytic semigroups by strongly elliptic operators under general boundary conditions, American Mathematical Society, Vol. 259, No. 1, pp. 299-310. 77

[11] Bucur A. (2006) About the second-order equation with variable coefficients, *General Mathematics*, Vol. 14, No. 3, p. 39-42 27, 43

[12] Cannon J. R. (1968) Determination of an unknown heat source from overspecified boundary data, *SIAM J. Numer. Anal.*, Vol. **5**, p. 275-286. 25, 40, 65, 100

[13] Cox B. A. (2003) A review of currently available in-stream water-quality models and their applicability for simulating dissolved oxygen in lowland rivers, The Science of the Total Environment 314-316, pp. 335-377 38, 67, 100, 101

[14] Dahlberg B., Trubowitz E. (1984) The Inverse Sturm-Liouville Problem III. Comm. Pure Appl. Math., 37, p. 255-267 42, 52

[15] Dobbins, W. E (1964) "BOD and oxygen Relationships in treams" J. A. Sce. San Div. 90(3) :53 18, 19

[16] El Badia A., Ha Duong T (2002) On an inverse source problem for the heat equation. Application to a pollution detection problem, Inverse Ill-Posed Problems 10 585–599 21, 25, 70

[17] El Badia A., Ha Duong T., Hamdi A. (2005) Identification of a point source in a linear advection dispersion reaction equation : application to a pollution source problem, *Inverse Problems*, Vol. **21**, Number 3, p. 1121-1136 25, 40, 42, 48, 65

[18] El Badia A., Farah M., (2006) Identification of dipole sources in an elliptic equation from boundary measurements, J. Inv. Ill-Posed Problems 14 (4), 331-353. 10

[19] El Badia A. and Hamdi A. (2007) Inverse source problem in an advection dispersion reaction system : application to water pollution, *Inverse Problems*, Vol. **23**, Number 5, p. 2101-2120 25, 40, 42

[20] El Jai A. and Pritchard G. (1988) Sensors and controls in the analysis of distributed systems. New York ; Wiley 42

Références bibliographiques

[21] Engl H. W., Scherzer O. and Yamamato M.(1994) "Uniqueness of forcing terms in linear partial differential equations with overspecified boundary data", *Inverse Problems*, **10**, p. 1253-1276. 40, 65, 100

[22] Farah M. (2006) Problèmes Inverses de sources et lien avec l'electro-Encéphalo-Graphie, Université de Compiègne (Thèse de doctorat) 10

[23] Fardin Boustani, Mohammah Hosein Hojati (2010) Pollution and Water Quality of the Beshar River, World Academy of Science Engineering and Technology, Issue 70 : 97-101 23, 38, 100

[24] Fischer H B, List E G, Koh R C Y, Imberger J & Brooks N H (1979) Mixing in Inland and Coastal Waters, Academic Press, New York, NY. 114

[25] George L. Bowie, william B. mills(1985) "Rates, constants and kinetics Formulations in Surface Water Quality Modeling" (Environmental Research Loboratory Office of Research and Development U.S Environmental Protection Agency, Athens, Georgia 30613). 15, 17

[26] Gerard T., J.Wiley (1982) "Mathematical Modeling of water Quality : Streams, Lakes and Reservoirs" Masson. 16, 17

[27] Hadamard J.. (1923) Lectures on Cauchy's Problem in Linear Partial Differential Equations. Yale University Press. 9

[28] Hamdi A. (2005) Identification de sources de pollution dans les eaux de surface, Université de Compiègne (Thèse de doctorat) 15, 16

[29] Hamdi A. (2007) Identification of point sources in two dimensional advection-diffusion-reaction equation : application to pollution sources in a river. Stationary case, *Inverse Problems in Science and Engineering*, Vol. **15**, Number 8, p. 855-870 25

[30] Hamdi A. (2009) The recovery of a time-dependent point source in a linear transport equation : application to surface water pollution, Inverse Problems, Volume 25, Number 7, p. 75006-75023 25, 41, 46, 65

[31] Hamdi A. (2009) Identification of a time-varying point source in a system of two coupled linear diffusion-advection-reaction equations : application to surface water pollution, Inverse Problems, Volume 25, Number 11, p. 115009-115029 25, 41

[32] Hamdi A., Mahfoudhi I. (2010) Boundary null-controllability of linear diffusion–reaction equations, C. R. Acad. Sci. Paris, Volume 348, Issues 19-20, pp 1083-1086.

Références bibliographiques

[33] Hamdi A. (2011) Inverse source problem in a 2D linear evolution transport equation : detection of pollution source, *Inverse Problems in Science and Engineering*, to appear. 25

[34] Hamdi A. (2012) Inverse source problem in a 2D linear evolution transport equation : detection of pollution sources, Inverse Problems in Science and Engineering, Vol. 20, No. 3, pp. 401-421 65, 66, 99, 118

[35] Hamdi A., Mahfoudhi I. (2013) Inverse source problem in a one-dimensional evolution linear transport equation with spatially varying coefficients : application to surface water pollution, Inverse Problems in Science & Engineering, to appear. 10, 37, 66, 100

[36] Hamdi A., Mahfoudhi I. (2013) Inverse source problem based on two dimensionless dispersion-current functions in 2D evolution transport equations with spatially varying coefficients, Inverse Problems in Sciences and Engineering, submitted. 10, 65, 100, 106, 123

[37] Hamdi A., Mahfoudhi I. and Rejaiba A. (2013) Identification of the time active limit with lower and upper frame bounds of the total amount loaded by an unknown source in a 2D transport equation, Inverse Problems and Imaging, submitted. 99

[38] Hettlich F. and Rundell W. (2001) Identification of a discontinuous source in the heat equation, *Inverse Problems*, Vol. **17**, p. 1465-1482 25, 40, 65, 100

[39] Kernévez J.P. (1997) "The Sentinel Method and Its Application to Environmental Pollution Problems" Boca Raton, New York. 16

[40] Koketsu K (2000) *Inverse Problems. Inverse Problems in Seismology*, Bulletin of the Japan Society for Industrial and Applied Mathematics, Vol. 10, Number 2, P. 110-120. 10, 37

[41] Kurbanov V. M., Safarov R. A. (2003) On uniform convergence of orthogonal expansions in eigenfunctions of Sturm-Liouville operator, Transactions of NAS of Azerbaijan, p. 161-168 42

[42] Lin F. H. (1990) A uniqueness theorem for parabolic equations, Comm. Pure Appl. Math., Vol. 43, pp. 125-36, MR 90j :35106 77, 79, 108

[43] Linfield C et al (1987) The enhanced stream water quality models QUAL2E and QUAL2E-UNCAS : *Documentation and user manual, EPA :* 600/3-87/007. 24, 38, 67

[44] Ling L., Yamamoto M., Hon Y. C. and Takeuchi T. (2006) Identification of source locations in two-dimensional heat equations, *Inverse Problems*, Vol. **22**, p. 1289-1305 25

[45] Lions J. L. (1988) Contrôlabilité Exacte Pertubations et Stabilisation de Systèmes Distribués, Tome 1 : Contrôlabilité Exacte, volume 8 of Recherches en Mathématiques Appliquées. Masson. 76, 106

[46] Lions J L (1988) Exact controllability, stabilization and pertubations for distributed systems. SIAM Review, 30(1) :1-68. 76, 106

[47] Lions J.L. (1992) Pointwise control for Distributed Systems in Control and Estimation in distributed Parameters Systems, Edited by Banks H.T. SIAM. 25, 40, 69, 77

[48] Myung E L and IL Won S (2010) 2D Finite Element Pollutant Transport Model for Accidental Mass Release in Rivers, KSCE Journal of Civil Engineering, 14(1) pp 77-86. 18, 67, 68, 101, 102

[49] Oelkers E. H. (1996). Physical and chemical properties of rocks and fluids for chemical mass transport calculations in Reactive Transport in Porous Media, P. C. Lichtner, C. C. Steefel and E. H. Oelkers, eds., The Mineralogical Society of America, 131–191. 18, 67, 68, 101, 102

[50] Okubo (1980) Diffusion and ecological problems : mathematical models, *Springer-Verlag, New York.* 24, 38, 67

[51] Pérez Guerrero J. S., Skaggs T. H. (2010) Analytic solution for one-dimensional advection-dispersion transport equation with distance-dependent coefficients, Journal of Hydrology, Vol. 390, No 1-2[Amsterdam ; New York] : Elsevier, p. 57-65 33, 57

[52] Rasmussen J M (2004) Boundary Control of Linear Evolution PDEs-Continuous and Discrete, Ph. D. Thesis, Technical University of Denmark. 85, 100, 106

[53] Rundell W., Sacks P. (1992) Reconstruction Techniques for Classical Inverse Sturm-Liouville Problems, Mathematics of Computation, Vol. 58, Num. 197, p. 161-183 52

[54] Rauch W., M. Henze, L. Koncsos, P. Shanahan, L. Somlyody and P. Vanrolleghem (1998) "River water quality modelling : I. state of the art" IAWQ Biennial INternational Conference, Canada. 16, 17

[55] Sakamoto K. and Yamamoto (2012) Inverse source problem with a final overdetermination for a fractional diffusion equation, Mathematical control and related fields, Vol. 1, Num. 4, pp. 509-518

[56] Saut J. C. and Scheurer B. (1987) Unique continuation for some evolution equations, J. Differential Equations, 66(1) pp. 118-139

[57] Schwartz L. (1966) "Théorie des distributions" Hermann paris. 25, 58

[58] Simon J (1983) *Caractérisation d'un espace fonctionel intervenant en contrôle optimal, Annales Faculté des Sciences Toulouse, Vol. V, pp 149-169.* 65, 70

[59] Solodky S.G. and Mosentsova A. (2008) Morozov's discrepancy principle for the Tikhonov regularization of exponentially ill-posed problems, Computational Methods in Applied Mathematics, Vol. 8, Num. 1, p. 86-98 *51*

[60] Titchmarsh, E. C. (1937) Introducton to the theory of Fourier Integrals, Oxford : Clarendon Press; *48, 80*

[61] Veerle Ledoux (2007) Study of Special Algorithms for solving Sturm-Liouville and Schroedinger Equations, Ph.D. Thesis, Ghent University Belgium. *56*

[62] Whitney M. L. (2009) Theoretical and Numerical study of Tikhonov's regularization and Morozov's discrepancy principle, Mathematics Theses : Georgia State University *51*

[63] Wikipedia *17*

[64] Xanthis C G, Bonovas P M and Kyriacou A G (2007) Inverse problem of ECG for different equivalent cardiac sources, PIERS Online, Vol.3, No. 8. 1222-1227 *10, 37*

[65] Yamamoto M. (1993) Conditional stability in determination of force terms of heat equations in a rectangle, Mathl. Comput. Modelling Vol. Vol. 18, Num. 1, p. 79-88 *40*

[66] Yamamoto M (1994) Conditional stability in determination of densities of heat sources in a bounded domain, International Series of Numerical Mathematics, Vol. **18**, pp. 359-370, Birkhauser, Verlag Basel. *40, 65, 100*

Résumé

Ce livre porte sur l'étude de quelques questions liées à l'identifiabilité et l'identification d'un problème inverse non-linéaire de source. Il s'agit de l'identification d'une source ponctuelle dépendante du temps constituant le second membre d'une équation de type advection-dispersion-réaction à coefficents variables. Dans le cas monodimensionnel, la souplesse du modèle stationnaire nous a permis de développer des réponses théoriques concernant le nombre des capteurs nécessaires et leurs emplacements permettant d'identifier la source recherchée d'une façon unique. Ces résultats nous ont beaucoup aidés à définir la ligne de conduite à suivre afin d'apporter des réponses similaires pour le modèle transitoire. Quant au modèle bidimensionnel transitoire, en utilisant quelques résultats de nulle contrôlabilité frontière et des mesures de l'état sur la frontière sortie et de son flux sur la frontière entrée du domaine étudié, nous avons établi un théorème d'identifiabilité et une méthode d'identification permettant de localiser les deux coordonnées de la position de la source recherchée comme étant l'unique solution d'un système non-linéaire de deux équations, et de transformer l'identification de sa fonction de débit en la résolution d'un problème de déconvolution. La dernière partie de ce livre discute la difficulté principale rencontrée dans ce genre de problèmes inverses à savoir la non identifiabilité d'une source dans sa forme abstraite, propose une alternative permettant de surmonter cette difficulté dans le cas particulier où le but est d'identifier le temps limite à partir duquel la source impliquée a cessé d'émettre, et donc ouvre la porte sur de nouveaux horizons.

Mots clés : Problèms Inverse de source ; Nulle contrôlabilité frontière ; Optimisation ; équation de diffusion-Advection-réaction ; Équation aux Dérivées Partielles ; Pollution des eaux de surfaces.

Abstract

The book deals with the two main issues identifiability and identification related to a nonlinear inverse source problem. This problem consists in the identification of a time-dependent point source occurring in the right hand-side of an advection-dispersion-reaction equation with spatially varying coefficients. Starting from the stationnary case in the one-dimensional model, we derived theoritical results defining the necessary number of sensors and their positions that enable to uniquely determine the sought source. Those results gave us a good visibility on how to proceed in order to obtain similar results for the time-dependent (evolution) case. As far as the two-dimensional evolution model is concerned, using some boundary null controllability results and the records of the generated state on the inflow boundary and its flux on the outflow boundary of the monitored domain, we established a constructive identifiability theorem as well as an identification method that localizes the two coordinates of the sought source position as the unique solution of a nonlinear system of two equations and transforms the identification of its time-dependent intensity function into solving a deconvolution problem. The last part of this thesis highlights the main difficulty encountred in such inverse problems namely the non-identifiability of a source in its abstract form, proposes a method that enables to overcome this difficulty in the particular case where the aim is to identify the time active limit of the involved source. And thus, this last part opens doors on new horizons and prospects.

Keywords : Inverse source problem ; Null boundary controllability ; Optimization ; Advection-Dispersion-Reaction equation ; Partial Differential Equation ; Surface water pollution.

i want morebooks!

Buy your books fast and straightforward online - at one of world's fastest growing online book stores! Environmentally sound due to Print-on-Demand technologies.

Buy your books online at
www.get-morebooks.com

Kaufen Sie Ihre Bücher schnell und unkompliziert online – auf einer der am schnellsten wachsenden Buchhandelsplattformen weltweit! Dank Print-On-Demand umwelt- und ressourcenschonend produziert.

Bücher schneller online kaufen
www.morebooks.de

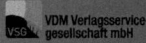 VDM Verlagsservicegesellschaft mbH
Heinrich-Böcking-Str. 6-8   Telefon: +49 681 3720 174   info@vdm-vsg.de
D - 66121 Saarbrücken   Telefax: +49 681 3720 1749   www.vdm-vsg.de

Printed by Books on Demand GmbH, Norderstedt / Germany